ADVANCED POLYMER PROCESSING OPERATIONS

ADVANCED POLYMER
PROCESSING
OPERATIONS

Edited by

Nicholas P. Cheremisinoff, Ph.D.

NOYES PUBLICATIONS
Westwood, New Jersey, U.S.A.

Copyright © 1998 by Nicholas P. Cheremisinoff
No part of this book may be reproduced or utilized in
any form or by any means, electronic or mechanical,
including photocopying, recording or by any information storage and retrieval system, without permission
in writing from the Publisher.
Library of Congress Catalog Card Number: 97-51237
ISBN: 0-8155-1426-3
Printed in the United States

Published in the United States of America by
Noyes Publications
369 Fairview Avenue
Westwood, New Jersey 07675

10 9 8 7 6 5 4 3 2 1

Library of Congress Cataloging-in-Publication Data

Advanced polymer processing operations / edited by Nicholas P.
    Cheremisinoff.
        p.   cm.
    Includes bibliographical references and index.
    ISBN 0-8155-1426-3
    1. Polymers.   I. Cheremisinoff, Nicholas P.
TP1087.A38   1998
668.9--dc21   97-51237
                                                             CIP

PREFACE

This volume is part of a series being developed by Noyes Publishers on applied polymer science and technology. The series being developed is designed to provide state–of–the–art design and technology information on polymers for engineers, product development and applications specialists, and end users of these materials. This volume covers advanced polymer processing operations and is designed to provide a description of some of the latest industry developments for unique products and fabrication methods. Contributors for this volume are from both industry and academia from the international community. This book contains nine chapters covering advanced processing applications and technologies. Subject areas covered include the processing of unsaturated polyesters and various prepolymers, new PVC processing techniques, PES and Nylon–3 chemistry, applications and processing methods, reactive extrusion technologies, latest developments and applications of pultrusion processing operations, electron beam processing of polymers, latest developments in the processing of thermoplastic composites, and the application of polymer technology to metal injection molding.

This volume and subsequent ones are geared toward industry applications, and as such emphasize commercialization aspects and industrial operations. The editor extends a heartfelt thanks to the contributors of this volume, and a special thanks to Noyes Publishers for the fine production of this volume.

Nicholas P. Cheremisinoff, Ph.D.
Editor

NOTICE

To the best of our knowledge the information in this publication is accurate; however, the Publisher does not assume any responsibility or liability for the accuracy or completeness of, or consequences arising from, such information. This book is intended for informational purposes only. Mention of trade names or commercial products does not constitute endorsement or recommendation for use by the Publisher. Final determination of the suitability of any information or product for use contemplated by any user, and the manner of that use, is the sole responsibility of the user. We recommend that anyone intending to rely on any recommendations of materials or procedures mentioned in this publication should satisfy himself as to such suitability, and that he can meet all applicable safety and health standards.

ABOUT THE EDITOR

Nicholas P. Cheremisinoff heads the Industrial Waste Management Program to eastern Ukraine under the United States Agency for International Development. He has nearly twenty years of industry and applied research experience in polymers, petrochemicals, and environmental and energy management in the heavy manufacturing and processing industries. Among his experience includes nearly thirteen years as product development manager and specialist for Exxon Chemical Company's elastomers product lines, and he actively provides consulting for private industry in the polymer technology areas. He has contributed extensively to the industrial press by having authored, co–authored or edited over 100 reference books and numerous articles. Dr. Cheremisinoff received his B.S., M.S. and Ph.D. degrees in chemical engineering from Clarkson College of Technology, Potsdam, New York.

LIST OF CONTRIBUTORS

Anil K. Bhowmick. Rubber Technology Centre, Indian Institute of Tecnology, Kharagpur 721 302 W.B. India

Tulin Bilgic. Petkim Petrochemicals Research Center, P.O. Box 9, 41740 Korfez/Kocaeli/Turkey

Tapan K. Chaki. Rubber Technology Centre, Indian Institute of Tecnology, Kharagpur 721 302 W.B. India

Sujit K. Datta. Rubber Technology Centre, Indian Institute of Tecnology, Kharagpur 721 302 W.B. India

M.L. Foong. School of Mechanical and Production Engineering, Nanyang Technological University, Nanyang Avenue, Singapore 639–798

Gungor Gunduz. Kimya Muhendisligi Bolumu, Orta Dogu Teknik, Universitesi, Ankara 06531, Turkiye

Rui Huang. Department of Plastics Engineering, Chengdu University of Science and Technology, Chengdu 610065, Sichuan, P.R. China

Viera Khunova. Department of Plastics and Rubber, Faculty of Chemical Technology, Slovak Technical University, Slovak Republic

Junzo Masamoto, Research Fellow of Asahi Chemical and Visiting Professor of Kyoto University. Polymer Development Laboratory, Asahi Chemical Industry Co., Ltd., 3–13, Ushiodori, Kurashiki 712 Japan

S.M. Moschiar. Institute of Materials Science and Technology (INTEMA) Juan B. Justo 4302. 7600 Mar del Plata, Argentina

M.M. Reboredo. Institute of Materials Science and Technology (INTEMA) Juan B. Justo 4302. 7600 Mar del Plata, Argentina

M.M. Sain. Pulp and Paper Research Center, University of Quebec, Trois–Rivieres, PQ, Canada

K.C. Tam. School of Mechanical and Production Engineering, Nanyang Technological University, Nanyang Avenue, Singapore 639–798

Kun Qi. Department of Materials Science and Engineering, Guangdong University of Technology, Guangzhou 510090, P.R. China

A. Vazquez. Institute of Materials Science and Technology (INTEMA) Juan B. Justo 4302. 7600 Mar del Plata, Argentina

CONTENTS

1

Processing of Unsaturated Polyesters

Güngör Gündüz
Orta Doğu Teknik Üniversitesi; Turkey

INTRODUCTION

Unsaturated polyesters (UPs) are composed of prepolymers with some unsaturation on the backbone and a vinyl monomer. The prepolymer is obtained from the interaction of saturated and unsaturated dibasic acids with dihydric alcohols. The vinyl monomer crosslinks the unsaturated acid moieties on the backbone upon polymerization, and a three-dimensional network is formed. They can be formulated to be hard and brittle, or soft and flexible depending on the chemical structure of acids, alcohols and monomers.

The history of UPs began just before World War II[1-3] and it was commercialized in 1941. Shortly after WW II fiber reinforced UP products appeared in the market. UP resin systems were the first widely used fiber reinforced materials in industry and they gave a high impetus and acceleration to the growth of a new field in industry. UP resins since then have been the workhorse of polymer composites.

They offer high mechanical strength, with a high strength-to-weight ratio, good chemical resistance, electrical insulation, low cost, ease of handling, and very high versatility.

1

RESINS

Chemical Composition

For dihydric alcohols, propylene, ethylene, and diethylene glycols and for acids-phthalic, adipic, and maleic anhydrides are commonly used. The use of maleic anhydride (or acid) is a necessity to incorporate available sites on the prepolymer backbone for the interaction with vinyl monomers. The most commonly used vinyl monomer is styrene while α-methyl styrene, methyl acrylate, methyl methacrylate, acrylonitrile, diallyl phthalate, and triallyl cyanurate can be used as comonomers.

The production technique is quite well known [4-6] and no sophisticated equipment or controls are needed. The reaction is carried under an inert atmosphere in a jacketed batch reactor equipped with a stirrer and cooler. The resin produced has a pale straw color mainly due to hydroquinone added as an inhibitor, and its viscosity is such that it resembles honey.

UPs can be formulated to be brittle and hard, or soft and flexible. Propylene glycol (1,3) produces a hard product while relatively long alcohols such as diethylene glycol gives high flexibility to polymer chains and thus to final products. Propylene glycol and diethylene glycol can be mixed at proper ratios to synthesize a polymer of desired mechanical properties. The rigidity of the product can be changed also by changing the ratio of saturated/unsaturated acids. The increase in the ratio increases flexibility.

The resin produced from propylene glycol, orthophthalic anhydride, maleic anhydride and styrene has a dominant use in many applications due to its inexpensive price though it has limited thermal stability, and chemical resistance. Isophthalic acids give products of higher quality, with better mechanical, thermal and chemical properties than the corresponding ortho resin products, but they are relatively expensive.

The use of bisphenol A fumarates introduces aromaticity into the structure and products with high thermal stability, chemical resistance and hardness are obtained. These products are exclusively used in high performance applications.

The partial substitution of styrene by acrylates lowers the viscosity, and provides better adhesion to the fiber in composites. Acrylonitrile imparts exceptional mechanical properties and increases both hardness and impact strength. The propylene glycol based UP with 40 % styrene has an impact strength of 14 J/m-width, and an addition of 20 % acrylonitrile increases it to 39 J/m-width. The addition of 11 % acrylonitrile increases hardness from 12 BUN to 26 BUN [7].

Some halogenic compounds like tetrachloro- or tetrabromophthalic anhydride can be used in the synthesis of prepolymer to make flame retardant resins. The bromine content must be around 12 % by weight to make a self-extinguishing polyester [8]. Flame retardant compounds either chemically attached to the backbone or physically added to the resin have a tendency to lower the mechanical strengths.

Polystyrene chains formed upon polymerization connect maleic anhydride moieties on the prepolymer chains and thus form a three dimensional network. The maleic anhydride transforms into fumarate in the course of polyesterification at high temperature. This transformation must be accomplished during the synthesis of prepolymer since fumarate has higher reactivity than maleate for the reaction with styrene.

Another type of prepolymer is obtained by the interaction of a monofunctional unsaturated acid with a bisphenol diepoxide having unsaturated sites at the two ends of the chain. It is then mixed with a vinyl monomer such as styrene. This resin is called vinyl ester, and its appearance, handling properties, and cure are similar to UP resins. They cost more than UPs but have exceptional mechanical and chemical properties.

Besides these conventional resins, UP based interpenetrating polymers, and isocyanate thickened UP resins [9,10] seem to have promising importance in the future.

Forming any item in the final shape becomes possible through the hardening (or curing) of UP resin. Peroxides, and azo and azine compounds can be used as initiator (or catalyst) at an amount of 1 - 2 %. The curing temperature is fixed by the decomposition temperature of the initiator. For room temperature applications methyl ethyl ketone peroxide, for moderate temperature (~ 60 - 90 C) benzoyl peroxide, and for hot press or oven curing (~ 130 - 150 C) benzoyl peroxide mixed with di-t-butyl peroxide or t-butyl perbenzoate is used at an amount of a few percent of the resin. To accelerate the decomposition of peroxides, cobalt naphthenate or cobalt octanoate can be used at quite low quantities, about 0.01 %. The excess amount of accelerator causes darkening of color and bubble formations inside the products. The peak exotherm has to be controlled to obtain a good quality product.

Choice of Resin

General purpose resin is synthesized from 1,3 propylene glycol, phthalic anhydride, maleic anhydride and styrene. Phthalic and maleic anhydrides are at equimolar quantities. This resin is the most available one in the market and suitable for most of the cold-set lay-up work and some hot molding. The substitution of some maleic anhydride by adipic or sebacic acid gives flexible products. Similar property resins can be obtained also by partial substitution of propylene glycol by diethylene glycol. These resins are sometimes called plasticized resins. The items made from these resins are relatively soft, have high impact strength but low flexural and tensile strengths. Acrylonitrile may be added to improve the mechanical properties.

These resins may be sold separately and they can also be blended with general purpose resins to have the desired properties.

In the production of fiber reinforced objects, a quick setting resin is applied to the surface of a mold and gelled before lay-up. This is called gel-coat and it improves the surface appearance of the laminate. Plasticized UP resins are often preferred to prepare gel coats. The flexible surface layer can easily be an integral part of the finished laminate.

For tough and heat resistant duties, the resins with high aromaticity must be preferred. Additives and fillers incorporate significant property changes, so they must be added at the optimum quantities.

Safety

UP resins must be kept below 25 C to have reasonable shelf-life. Heat, sunlight, and energetic radiation affect the shelf-life adversely. UP resins are classified as inflammable materials due to styrene in the composition. Adequate ventilation is needed in the store rooms.

Peroxides are powerful oxidizing agents and must be handled with care. Initiators and accelerators must be kept in a cool and dark place to minimize decomposition. Accelerators should not be mixed with initiators, otherwise a violent reaction may occur. Peroxides must be thoroughly stirred with resins before adding accelerators.

CASTING

UPs have been used extensively as cold-setting potting compounds in diversified applications. There is no restriction on the size and shape of the object to be produced. Large and intricate items can be successfully produced with fineness of detail in inexpensive molds. The initiator and accelerator are first added homogeneously to the resin which is then poured into the mold.

UP casting was used in the past to produce decorative items, pearl buttons, knife and umbrella handles in attractive pastel shades, and to encapsulate parts and assemblies in the electronic industry. The most important casting application has been the manufacture of pearl buttons. The resin mixed with suitable pigments is cast into a thin sheet in a cell kept upright. If organic dyestuffs are used as pigments, they should not decolorize due to temperature rise during gelation of the resin. Anatase titanium dioxide and carbon black should not be preferred as white and black colorants because they inhibit the cure of the resin. The buttons are stamped out of the cast sheet before it is fully hardened. Centrifugal casting machines can also be used to produce sheets. In this case a fast cold-setting catalyst system, must be used or the curing temperature must be raised to 60 C. After the buttons have been cut out, they are post-cured preferably in hot glycerine. The button blanks produced are then machined.

Automation in the casting process is usually impractical because most of the production is in the line of decorative reproduction of an already existing valuable object. However centrifugal and rotational casting devices can be used to mold hollow shapes. The speed usually changes between 10 - 100 rpm [11].

Porous objects such as plaster and concrete can be strengthened by UP impregnation. However vinyl monomer impregnation is often substituted for UP impregnation since vinyl monomers have much lower viscosities and they penetrate into much smaller pores [12]. Strengthening of the loose archeological objects, or adhesion of broken parts can be successfully accomplished by UP resins.

COMPOUNDING MATERIALS

UP resins can be compounded with different additives and filling materials to improve and enhance the physical and mechanical properties. The added material must be uniformly distributed within the resin. The compound is then molded to the desired shape by several techniques such as bulk molding, transfer molding, or sheet molding.

Additives

Chemicals and nonreactive materials can be added to resin formulations to achieve particular properties in the final products. The type and amount of additives have predominant influences on the shape and quality of the final products [13].

Fillers

Fillers are usually inorganic inert materials in powder or fiber form, and are usually added to reduce cost. They improve stiffness by improving modulus.

Fillers must be carefully selected to achieve the required property of the product. Several fillers can be combined at the desired proportions to impart their individual properties to the product. Fine grinding increases surface to volume ratio which in turn increases the resin requirement and thus the cost. In general particle size must not fall below 200 mesh except for coating application.

Fillers must be as clean as possible and free of oily materials, dirt, dust and especially moisture. Moisture exhibits complicated problems in compression-molding. It may cause partial polymerization and defects such as pinholes and pores especially in cast products. The pH of fillers and the acid number of resin must be such that excessive weakening should not happen at the interface.

Fillers usually comprise 10 - 90 % of the total weight of the mix, and large amounts are usually referred as extenders. They lower the cost and also the mechanical properties, so the amount of fillers must be kept at such levels which give the required mechanical strengths.

Powdered Fillers

The addition of powdered fillers at optimum quantities increases the compression strength. Excessive amounts decrease the flexural and tensile strengths. The volumetric contraction in the resin after curing is also reduced by the presence of fillers. In addition they retard the flow of resin in hot-molding.

Woodflour obtained from hardwoods or nut shells is one of the most widely used filler. It is cheap, readily available, and strong due to its fibrous nature. In addition it can be easily wet by resin, and hence, can be readily compounded. However the moisture existing in the woodflour creates some problems and gives poor electrical properties and low dimensional stability. Sawdust, wood pulp, jute and other cellulosic materials can also be used as fillers.

Fillers of mineral origin are used for a variety of purposes to affect physical, mechanical, electrical properties, and the appearance. Almost every crushed and ground rock may be compounded with UP resins. Hard carbonates such as calcium carbonate, nonreactive sulphates such as barium sulphate (baryte), and some metal oxides are used as fillers, and they result in a white color compound. Silica, ceramic oxides, diatomaceous earth and asbestos exhibit thermal and electrical insulation. Mica flour exhibits good electrical properties but poor heat insulation. The use of asbestos is highly restricted due to its health effects on the human respiratory system. The properties exhibited by some well-known fillers are given below [14].

Calcite (or calcium carbonate):	Inexpensive and most available filler; most widely used.
Clays:	Low cost fillers mostly used in dough molding: impart plasticity similar to dough.
Dolomite:	Used in controlling the degree of wetness in dough molding due to its low oil absorption property.
Silica and alumina:	Provides hardness and improves abrasion resistance, and thermal insulation.
Mica flour:	Improves electrical insulation.
Expanded vermiculate, pearlite or pumice:	Used to make lightweight products.

Chemical resistance is improved by glass fibers, synthetic fibers, metal oxides, and graphite. The effects of different fillers on physical and mechanical properties of products are briefly given in Table 1[15].

Effect of Moisture

Crushed or ground silica, quartz, granite, and baryte exhibit minimum water absorption while clays, wood flour, and other cellulosic materials are the most absorptive among all fillers. Calcium sulphate (i.e., gypsum) can not be used in compounding UP resins due to its two moles of water of which one and half moles evaporate above 120 C. Moisture becomes a real problem when fillers are used at large percentages. The resin must undergo gelation at the right time with the right amounts of initiator and accelerator. Gelation is decreased 20 % by 1 % water, 30 % by 2 % water, and 40 % by 3 % water. Moisture contents above 2 % result in premature gelation and the cured resin exhibits loose structure and breaks apart. The presence of water also causes shrinkage in the hardened resin. The widely used phthalic anhydride based resin shrinks 3.25 % by 1% water, and 3.80 % by 2 % water. Above 0.5 % water content, poor surface finish results and the release from the mold becomes difficult. If the water content exceeds 1 % the hardened product shows poor weather and chemical properties, and likewise the mechanical properties are lowered by as much as 40 % [16].

Table 1. Fillers and aggregates; properties and effects imparted to polyesters [from Ref. 15].

Legend:
L-low, E-excellent; M-medium, X-not recommended; H-high, O-no known effect; P-poor, A-asset; F-fair, D-detriment; G-good

Note: The columns *Shrinkage*, *Flexural modulus*, and *Elongation* carry the notations "All reduce shrinkage, in relation to absorption and %", "All reduce modulus, in relation to absorption and %", and "All reduce elongation" respectively (no individual entries).

Filler	Cost	Specific gravity	Resin absorption	Reactivity	Exterior exposure	Abrasion resistance	Shrinkage	Tensile strength	Compressive strength	Flexural modulus	Flexural strength	Impact strength	Elongation	Hardness (Barcol)	Dielectric properties	Electrical conductivity	Water resistance	Heat resistance	Flame resistance	Organic-acid resistance	Inorganic-acid resistance	Alkali resistance	Salt resistance	Sulfides resistance	Thermal conductivity	Thixotropy to resin	Pigmenting effect
Ground marble CaCO3	L	M	M	L	G	F		F	F		F			F	F		F	G	G	P	P	G	G	G	M	M	M
Marble chips	L	M	M	L	G	F		F	F		F			F	F		F	G	G	P	P	G	G	G	M	M	L
Dolomite	L	M	L	L	G	F		F	F		F			F	F		F	G	G	P	P	G	G	G	M	M	L
Treated silicas	M	M	M	O	E	E		F	F		F			F	F		E	E	E	E	G	G	G	M	H	O	
Pearlite	L	L	H	O	E	P		P	P		F			F	F		P	E	E	E	E	G	G	G	L	H	O
Ground silicas 400-20 mesh	L	M	L	O	E	E		F	E		F			E	E		E	E	E	E	E	E	E	E	H	L	O
Quartz, crushed	L	M	L	O	E	E		F	E		F			E	E		E	E	E	E	E	E	E	E	H	L	O
Granite, crushed	L	M	L	O	E	E		F	E		F			E	E		E	E	F	E	E	E	E	E	H	L	O
Mica, muscovite	M	M	L	O	G	F		G	F		F			G	F		G	G	F	P	P	G	G	P	M	L	O
Mica, biotite	M	M	L	O	G	F		G	F		F			G	F	P	G	G	F	P	P	G	G	P	M	L	L
Mica, synthetic	H	M	L	O	G	F		G	F		F			G	F		E	G	E	F	P	P	G	G	P	M	L	O
Talc	L	M	M	O	G	P		F	F		F			F	F		F	F	F	P	P	F	G	G	L	M	L
Asbestine	L	M	H	O	G	F		F	F		F			G	F		F	G	F	P	P	F	G	G	L	M	L
China clay (kaolin)	L	M	H	L	G	P		F	F		F			G	F		F	G	F	P	P	G	G	G	L	H	L
BaSO4 (barytes)	M	H	L	O	E	P		F	F		F			F	F		G	F	F	F	G	G	G	G	M	L	M
Asbestos, shorts	M	M	H	O	G	F		G	G		E			F	F		E	G	P	P	G	G	G	G	L	H	L
Graphite	M	M	L	O	G	E		F	G		G			F	X	G	E	E	E	E	E	G	E	G	M	L	H
Bentonite	L	M	H	L	G	P		F	F		F			F	F		G	F	P	P	G	G	G	G	L	H	L
Aluminum oxide	H	H	L	O	G	E		G	E		G			E	X	G	E	E	E	P	P	G	F	G	H	L	O
Boron carbide	H	M	L	O	G	E		F	E		F			E	X	G	E	E	E			G	G	G	H	L	O
Wood flour	L	L	H	O	G	F		E	F		E			E	P		F	P	P	F	F	G	F	L	H	O	
Zinc dust	M	H	L	O	G	P		F	F		F			F	X	F	G	F	F	P	P	F	P	F	H	L	H
Magnetite	M	H	L	O	G	E		F	F		F			F	X	G	G	G	E	P	P	G	G	G	L	H	H
Zirconium silicate	M	M	L	O	E	E		F	E		F			E	X	F	E	E	P	F	E	G	G	H	L	L	L
Diatomaceous earth	M	M	H	L	G	P		F	F		F			G	P	F	F	G	G	P	F	F	G	L	H	M	
Rice hulls, ground	L	L	L	O	G	F		G	G		E			F	F	P	F	P	P	F	F	F	G	F	L	M	L
Bagasse, ground	L	L	L	O	G	F		E	G		E			F	P		F	P	P	F	F	F	G	F	L	M	L
Nut hulls, ground	L	L	L	O	G	G		G	G		E			F	P		F	P	P	F	F	F	G	F	L	M	L
Hydrated alumina	H	M	L	O	E	E		G	E		E			G	X	G	E	E	E	P	P	F	F	F	G	H	L	M

The moisture content of most fillers can be lowered down to 0.02 % upon drying. However clays and wood floors cannot be dried down to such a low moisture content. So they must be used in small quantities in the mix to minimize the adverse effects of moisture.

Low - Profile Additives

UP resins undergo volume shrinkage by about 7-10 % +upon curing. This causes warpage, wavy surfaces, and internal voids and cracks especially in fiber reinforced products, and a post-mold processing may be needed to get the desired surface finish. A good method of overcoming this problem is to add so called low-profile or low-shrink materials to the mix. Some thermoplastic polymers such as polyvinyl acetate may be used for this purpose in small quantities of about 3 - 5 % [17]. It is believed that when styrene is absorbed on a low-profile additive, it polymerizes at a slower rate than the bulk, and boils in the late stages due to increase of the temperature, and it thus creates an internal pressure and compensates for the shrinkage [18]. In addition microstress cracking takes place at the interface between the thermoplastic additive and the bulk polyester. This is claimed to make a large contribution to low-shrink behavior [19, 20].

The thermoplastic low-profile additives compatible with UP resins are polyvinyl acetate, thermoplastic polyester, acrylics, styrene copolymers, polyvinyl chloride and its copolymers, cellulose acetate butyrate, polycaprolactones, and polyethylene powder [21].

Thickeners

Bulk and sheet molding compounds must have a viscosity above 10^4 Pa.s (or 10^7 cp) in order to be able to mold them without having any problems such as fluid leakage from the mold. An ordinary UP resin has a viscosity of about 1.6×10^3 cp. The tremendous increase in viscosity can be achieved by using group IIA metal oxides and hydroxides, especially magnesium oxide (MgO) and calcium hydroxide ($Ca(OH)_2$). They connect the carboxylic groups of chains forming a kind of network leading an increase in viscosity [22, 23]. CaO does not cause thickening alone while 3.6 % MgO increases the viscosity to 10^7 cp and 5.0% $Ca(OH)_2$ to 3×10^6 cp by the end of a one-week period. The 3.8 % CaO + 2.9 % $Ca(OH)_2$ composition gives 2.5×10^7 cp and 2.5 % CaO + 1.8% MgO gives 4.2×10^7 cp in one-week. Pure MgO is also a powerful thickening agent and about 2 % of it can increase the viscosity to 13.6×10^7 cps in two weeks. To inhibit such large increases, maleic anhydride can be added at an amount of 1-3 %. Other anhydrides such as benzole acid anhydride, tetrahydrophthalic, hexahydrophthalic, and phthalic anhydride accelerate thickening [24]. A thickener is a final ingredient added, and the mix must be used within a reasonable time, otherwise slow polymerization and excessive thickening result in a hard cake that is improper for molding.

Thixotropic Agents

When molding large composite objects with sharp corners and inclined surfaces, a severe technical problem is faced with the drainage of the resin. Not only with the resin/fiber ratio change, causing mechanical weakness, but styrene will also evaporate easily from the thin resin surface yielding insufficient cure. To use high viscosity resins may decrease drainage, but then wetting of the fibers may be difficult. Poor wetting naturally decreases the interface strength resulting in poor mechanical strength. This problem can be solved by adding thixotropic agents to the resin. Since thixotropic materials are gel-like at rest but fluid when agitated, a UP resin containing a thixotropic agent as low as 2 % will not drain from the corners or inclined surfaces. Some well-known thixotropic agents are silica aerogel, bentonite clay, china clay, polyvinyl chloride powder, iron oxide (Fe_2O_3), chrome oxide (Cr_2O_3), and acicular zinc oxide (ZnO). All of these shorten shelf-life and cure rate of the resin. Silica aerogel is the least hazardous one so it is most widely used.

Pigments

Pigments used to color the products can be of inorganic or organic origin. They are added at quantities not exceeding 5 percent. Blending the pigments with the resin is usually a difficult process. The particles or inorganic pigments are sticky and therefore vigorous mixing is needed to avoid particle agglomeration. Organic pigments are fluffy and carry electrostatic charges. Dry blending with other additives is difficult and hence they should be added directly to the resin.

Wetting of pigment particles by resin is a serious problem. Some surface active agents can be used to ease wetting. Improper dispersions affect color shade. Excessive use increases the cost, and some pigments affect the shelf life. They may accelerate or inhibit gelation. Some pigments are supplied in the form of paste dispersions. They can be easily added to the mix without any agglomeration problems. The alkaline pigments such as iron colors inhibit gelation and increase cure time. Hence, the catalyst content must be increased to overcome this difficulty. Synthetic-pearl pigments such as umbers, siennas, and ochers also inhibit gelation. Calcium carbonates and acicular zinc oxide used as pigments exhibit slight inhibition. Carbon blacks and anatase titanium dioxide are acidic and show inhibition effects, and when they are used as pigments, the initiator content must be reduced accordingly.

Pigments in phthalate and phosphate esters should not be used since they plasticize the resin resulting in a reduction in hardness [27]. The compatibility of organic pigments with UP resins is important, otherwise pigments migrate to the surface and give rise to undesired surface properties.

Ultraviolet Absorbers

In places where prolonged exposure to sunlight is anticipated, UV absorbers must be added to the mix. They absorb the harmful UV radiation and dissipate in nonradiation form which is usually heat. The most well known UV absorbers are hydroxybenzophenones and hydroxyphenylbenzotriazoles [25]. UV stabilization is done by adding these compounds to the resin in the range of 0.1 - 0.25 %.

Flame Retardants

Inorganic hydroxides such as aluminium hydroxide and magnesium hydroxide decompose and give off water vapor on heating. Cooling by absorbing heat is the simplest technique for flame retardance. Some ammonia and sodium based boron compounds, and salts of phosphoric acid and of zinc and heavy metals give rise to the formation of a layer on the surface. The coating thus formed prevents the reaction between the resin and oxygen.

Chlorinated and brominated organic additives are known to be the best fire extinguishers. Bromine compounds though expensive are more effective than chlorine. Bromine is heavier than chlorine and more loosely bound to its molecule. On heating, it easily leaves its molecule and combines with the hydrogen radical to form hydrogen bromide or with other radicals formed from the decomposition of polymer chain. Hydrogen bromide is a powerful radical scavenger, it stops chain propagation during combustion. Iodine compounds cannot be used for flame retardance because iodine is very loosely bound to its molecule and not stable even at room temperature. Florine can not also be used for this purpose because it is too strongly bound to its molecule. The mechanism of flame retardance by halogenic compounds is quite complicated. However organic bromine compounds added to UP resins in such quantities to incorporate about 12 % bromine make them self-extinguishing.

The flame retardant additive must have high bromine content and a melting temperature higher than the softening temperature of UP resin. In recent years the use of decabromodiphenyl oxide (DBDPO) in a variety of resins has been quite popular. It has 83 % of bromine content, and has a melting point of 578 K. The percentage of oxygen in an oxygen + nitrogen mixture which burns the polymer is known as the oxygen index and its value is around 19 for UP resins. The oxygen index value must be increased above 25 to make a convenient self-extinguishing resin. This necessitates the consumption of large amounts of halogenated compounds in the resin. Synergetic chemicals such as antimony oxide (Sb_2O_3) may be used to decrease the need for halogenated compounds to increase the oxygen index value further. For instance 10 % DBDPO yields an oxygen index value of 21.80 % while 2 % Sb_2O_3 + 10 % DBDPO increases this value to 25.10 % of the propylene glycol based polyester [26]. Halogenated organic flame retardants either chemically bound to the backbone or physically added to the mix lower the mechanical properties of the product, so they are used only whenever absolutely needed.

Mold Release Agents

It is difficult to classify these materials because the types of molds or dies, temperatures, and conditions can vary so widely that the choice usually depends on experience and common sense. However, we can technically classify mold release agents as either internal or external depending on how they are applied in the process. Internally used agents are mixed into the resin and migrate to the surface on compression. Internal mold release agents may be used in quantities 0.25 -1 % of the resin.

Stearic acid and zinc stearate used as internal mold release agents may reduce the gloss of the finished product, while calcium stearate does not exhibit such adverse effect. Stearic acid should be used if the molding temperature is below 400°K. Zinc stearate has a melting point of 406°K and can be used up to 430°K, while calcium stearate melts at 423°K and can be used up to 440°K. At high temperature molding these compounds melt and form barrier at the mold-molding compound interface against adhesion. For high temperature applications fluorocarbons and some silicones can be successfully used as exterior release agents. Some refined soya oils, sodium or potassium alginates, and different waxes can be used as low temperature mold release agents.

The chemical structure of the resin has a predominant effect on its adhesion properties. Isophthalic, bisphenol A, and chlorostyrene increase the adhesion of the resin to the mold. The proper mold material must be selected for the type of the resin used.

REINFORCEMENTS

Fibers

Reinforced polyesters can be molded into extremely large shapes at atmospheric pressure or little pressure, and the products can be designed to provide practically any   shape. The reinforcing agent can be fibrous, powdered, spherical, or whisker made of organic, inorganic, metallic or ceramic materials. Fibrous reinforcements are usually glass, others such as asbestos, sisal, cotton are occasionally used. High modulus carbon, graphite, aramid or boron fibers are not preferred to reinforce UP resins.

Proper reinforcement materials can increase the strength several-fold. The principal reinforcement is glass fiber, and it accounts for 90 % of total usage. The most commonly used fiber is E-glass (electrical grade) which has very good dielectric properties, heat and flame resistance. It is an alumina-borosilicate with low alkali. The A-glass (high alkali) is mainly made of silica lime and soda and has good chemical resistance. ECR-glass contains mainly silica, alumina, and lime and owns good electrical properties and chemical resistance. S-glass (silica rich) is made of silica, alumina, and magnesia and exhibits high tensile strength and thermal stability. However it is relatively expensive and its use can be justified only under severe conditions. Filament fiber diameters change from 0.8 μm to 25 μm, and fibers are marketed in a variety of forms.

1. Continuous Roving

Fiberglass roving is produced by collecting a bundle of untwisted strands and wound into a cylindrical package. Continuous roving fibers are used in filament winding and pultrusion processes.

2. Chopped Strand

Continuous strands are chopped to desired lengths, typically 3 to 12 mm by a mechanical chopper. Screening is needed to eliminate improper material. Fibers can be sized by some adhesive polymers or resins to improve adhesion between fibers and the UP resin. The strands can be chopped in a wet state directly after sizing. Chopped fibers are mainly used in molding processes.

3. Woven Roving

Continuous roving can be woven to make products of different widths, thicknesses, weights, and strength orientations. Woven ravings exhibit high strength and rigidity, and used in lay-up processes to produce large size objects.

4. Woven Fabrics

Fabrics are made from yarns which are produced from twisted fine strands. Woven fabrics can easily handle strength orientations, and increase mechanical properties. They exhibit high strength biaxially, and good formability. They are used in wet lay-up and compression molding processes.

5. Mats

Chopped strand mats can be produced by randomly depositing chopped strands onto a belt and binding them with a polymer such polyvinyl acetate. It is partly softened and dissolved in the styrene monomer of UP resin. In fact the incorporation of polyvinyl acetate can be done at the production stage of fiber. Chopped strands have low formability, low washability, and low cost. So the mats made from them are used for making medium- strength objects with uniform cross-sections by compression molding and hand lay-up.

Continuous strand mat is formed from continuous strands with less binder requirement. They have good formability and wash resistance. They are used in closed mold processes and also in pultrusion where some transverse strength is required.

6. Combination Mats

These are comprised of alternate layers of mat and woven roving which are either bound by resinous binders, or stitched, or mechanically knit. They are used in the layup production of large parts. Figure 1 shows photographs of different types of fiber.

Fiber Sizing

Sizing may have both positive and negative effects on composite properties. Sizings are applied at quantities less than 1 %. The sizing materials used as film-formers are polyvinyl alcohol, polyvinyl acetate, starch and starch derivatives. Among this polyvinyl acetate shows the most satisfactory compatibility with UP resins. Silane coupling agents used as adhesion promoters enhance the interface strength. Silanes are often applied with a film-former material.

BULK MOLDING

In bulk molding processes, curing is achieved under elevated temperature and applied pressure. There are several possibilities.

Dough Molding

Resin mixed with initiators, and accelerators are blended with additives, fillers, and short fiber (i.e. chopped strand) reinforcements in a mechanical mixer such as sigma blade mixer. In case only fibrous powder is used as filler, it is recommended to add just a little nonfibrous powder to serve as a flow controller. There must exist sufficient blade clearance in the mixer. The blade speed is around 20 - 30 rpm, and adequate cooling may be needed. The filler and lubricant must be loaded before starting the blades. After mixing for a few minutes the resin containing initiator, accelerator, and pigments is added. Mixing is continued for about 10 - 20 minutes. The dough (or premix) is discharged from the mixer when the composition reaches the consistency of putty. The cake thus produced can be compression molded to

(a)

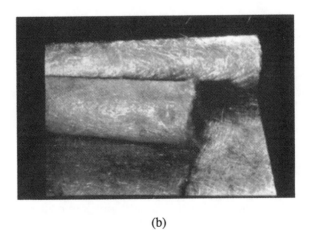

(b)

Figure 1. (a) Photograph of rovings;
(b) Photograph of chopped strand.

(c)

(d)

Figure 1 Continued: (c) Photograph of mats,
(d) Photograph of fabric (from Ref.14).

the desired shape as simply seen in Figure 2.

The mix containing benzoyl peroxide as initiator can be molded between 120 - 150 C under pressure for 10 to 20 minutes depending on the thickness of the item to be produced. A pressure around 15 MPa is satisfactory for most applications. Preforming of the cake may help to ease molding.

A large variety of objects can be produced by this technique. Since UP resins have good electrical properties, dough molding products find different applications in electrical field.

Preform Molding

This technique is also known as resin transfer molding and is especially suitable to produce complex shapes without high-cost tooling. It also provides mass production without size limitation.

A preform of the shape of the object is first made by depositing chopped glass fibers on a rotating perforated screen which has the general shape of the article to be molded. A suction is applied behind the screen to keep the fibers in place as shown in Figure 3. A binder which is usually polyvinyl acetate for UP resins is sprayed on chopped fibers. The preform is then taken from the screen and dried in an oven.

The preform thus produced is transferred to the lower part of a pair of heated matched metal dies, and a measured quantity of resin containing initiator and accelerator is poured or pumped on top. The press is then closed and compressed to a pressure of about 1.5MPa causing the resin to flow and wet out the preform. The molding time is about 1-5 min at 140°C.

Composites with fiber volumes more than 60 % can be made by this technique. It allows relatively low cycle times, good quality control, easily learned operator skills, and low capital investment. Typical articles produced by this technique are consumer products such as helmets and machine parts used at home, tanks, pipes, and automotive structures.

Injection Molding

Injection molding,which was once particularly suitable for thermoplastics, has found extended applications also in shaping properly formulated thermoset compounds. The molding compound is transported through runners and gates, and it flows into the mold cavity. It is essential that the compound must flow easily at lower-than-mold temperatures without curing. Meanwhile the components which are resin, filler, and fibrous glass must keep integrity and not separate during flow into the cavity. It is clear that long molecular weight or highly viscous resin, long reinforcing fibers or

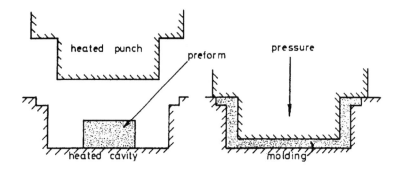

Figure 2. Principle of compression molding [from Ref.28].

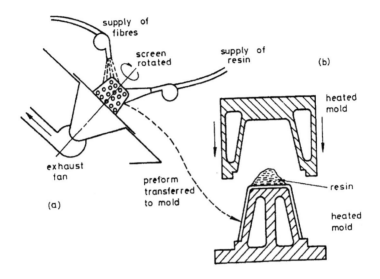

Figure 3. Preform molding [from Ref.28].

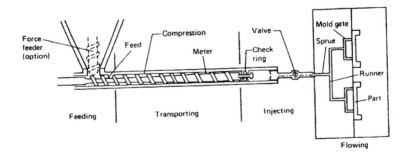

Figure 4. The injection molding system [from Ref.29].

flakes cannot be easily manipulated in injection molding. Despite these difficulties, items can be produced at high complexity with holes, ribs, and bosses.

There are several designs for the feed mechanism. The simplest one is the use of a plunger which pushes the material into the mold. However this gives rise to densification of the material. In another design the mix is forced from a cylindrical tank down to the entrance of a screw injection cylinder. The screw conveys the material to the mold at the front of the machine. Short screws are used for thermoset resins, and the compression ratio is below 1.5. Otherwise, heat may be produced and hardening may take place inside the machine. The barrel of the machine needs to be cooled to prevent premature polymerization. The typical temperatures at the rear and front of the cylinder are around 50 and 65°C respectively.

The usual speed of the screw is 20-75 rpm, and the injection pressure on material changes between 30 to 80 MPa. The residence time of the molding compound is 1-5 sec. A typical injection molding system is shown in Figure 4.

The molding compound is injected into the mold cavity of the object to be produced through a single hole known as a sprue. It then flows through the runners into the cavity. The runners must have large diameters and their length must be kept as short as possible, and sharp corners, restricted orifices and gates must be eliminated to have an easy flow of the compound. The sprue and runners can be water cooled to have a temperature of about 75°C at which some preheating can be achieved. The curing temperature in the mold is between 140 to 170°C.

Injection molding is used to fabricate automotive structural parts, power tool housings, garbage disposal units, and items with complex patterns.

SHEET MOLDING COMPOUNDS

In this process the molding compound is formed into a pliable sheet which then can be molded under pressure and heat. A typical machinery system used in the process is shown in Figure 5.

The resin paste containing the appropriate additives and fillers is fed on a plastic carrier film such as polyethylene or cellophane through a doctor plate. Glass fiber ravings are chopped and spread out on the paste layer. There exists another plastic carrier film beyond the chopping zone, and it brings another resin paste which covers the top of chopped ravings. The existing compaction cylinders sandwich the chopped rovings between two paste layers. Heating the cylinders aids wetting and starts thickening.

The sheet thus formed is continuously taken up and rolled on a turret.

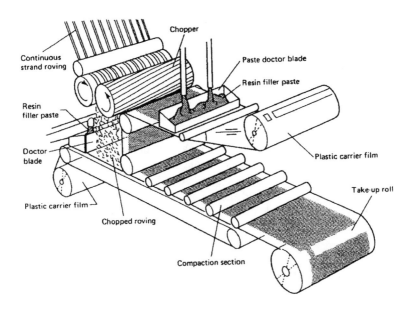

Figure 5. Sheet molding compound processing machine [from Ref.30].

When the appropriate weight is reached it is cut and taped for storage. The thickeners added to the resin paste provide sufficient high viscosity within a few days in the store room.

The sheet thus produced is a kind of prepreg made from chopped ravings and ready for molding. It is simply cut into smaller sizes and placed between the halves of a heated mold.

Chopped roving is used in the amounts ranging from 30 to 70 %. The sheet thus produced is suitable to produce items with varying cross sections. In addition to chopped roving, continuous glass fibers can also be used in making sheet molding compounds. Continuous glass fiber is directly fed to the system without having a chopper in Figure 5. This material has a high unidirectional strength. The resin impregnated sheet thus formed can be cut into pieces and wound on a mold surface in a traversing manner with certain angles, and then molded.

Sheets with both unidirectional fibers and chopped fibers can be produced by introducing a continuous strand roving feed mechanism to Figure 5 in between the chopper and the upper plastic carrier film. This provides higher strength with improved molding properties.

Continuous fibers can be wound on a large diameter cylinder in an X-pattern by using a standard filament winding machine. It is then cut, removed from the cylinder, flattened out and impregnated with resin. The sheets thus formed are compression molded and the products have quite high mechanical strengths. Chopped strands can be also incorporated into an X-pattern to improve the strength further.

WET LAY-UP PROCESSES

In these processes there is usually only one half of the mold, and the item to be produced comes out in the form of this mold. The resin has low viscosity without fillers and much additives, and it is spread on the continuous glass fibers stationed in the mold.

Hand Lay-Up

This is the oldest, and the simplest method of making polyester composites, and is still extensively used. The molds can be made from sheet metal, wood, plaster, or even composites.

The first step is to coat the mold with a mold release agent so that the cured item can be removed from the mold without damage. Fluorocarbons are preferably used for this purpose although silicones are used when the surface finish is not of primary importance. Silicones leave residues on surfaces that cause problems with subsequent bending, or painting. Polyvinyl alcohol can also be used as a mold release agent. The second step is to apply a gel-coat layer of approximately 0.3 - 0.4 mm thick to the surface. This gives a smooth surface finish and good appearance to the product, and prevents the moisture diffusion which weakens the resin-fiber interface.

After the gel-coat has partially cured an initial coat of resin is brushed on it and this is followed by laying up the reinforcements such as mat, woven roving, or fabric manually. The entrapped air is removed by hand rollers which also ensure complete fiber wet-out. Successive layers of reinforcements and resin can be alternatingly added to build up a desired thickness.

Curing takes place at room temperature with the use of a proper initiator such as methyl ethyl ketone peroxide.

There are several variations of the hand lay up technique, as shown in Figure 6.

Contact Molding

This is the basic hand lay-up process, and the resin is in contact with air. The wet lay-up hardens at room temperature. A smooth exposed surface can be achieved by covering the surface with a plastic film which can be removed after cure. Boat hulls, lorry cabs, or similar objects are molded in a female mold while water or chemical tanks must be molded over a male mold as a smooth inside surface is needed for them.

Complex patterns can be produced by using collapsing molds. The fiber is wound on a core made from paraffin or plaster. After cure, the paraffin can be melted out, and the plaster can be broken out.

Contact molding requires a minimum amount of equipment and presents no size restriction. The molds are low cost, and design flexibility enable a great variety of configurations such as boats, tanks, body components for the automotive industry, and housings.

Vacuum Bag Molding

Usually a female mold is used for ease of resin application. The lay-up prepared is covered with a flexible bag usually made from cellophane, or polyvinyl acetate.

The bag is clamped and vacuum is applied between the mold and the bag. It sucks the air and eliminates voids. Boats, boat components, and prototypes can be produced by this technique.

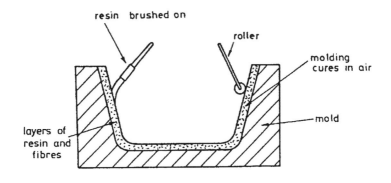

(a)   Basic hand lay-up method.

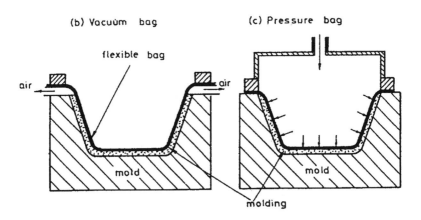

Figure 6. Hand-up techniques [from Ref.28].

Pressure Bag Molding

In this case a rubber sheet is used over the lay-up. Air pressure is applied up to 300-400 kPa. Cylindrical items, boats, safety helmets, containers, automotive or aircraft components, necked tanks, and large tubings can be produced by this technique.

SPRAY-UP PROCESS

In this process, chopped glass fiber roving and resin are simultaneously applied on a mold surface. The mold release agent, and the gel-coat are first applied on the surface.

Fiber glass roving is fed to a chopper, and the chopped strands are sprayed onto the mold simultaneously with resin containing initiator and accelerator as shown in Figure 7.

The mix material on the surface is rolled by hand to lay down fibers, remove air and to smooth the surface. The thickness is controlled by the operator and sufficient thickness can be built up in sections likely to be highly stressed. The success highly depends on the skill of the operator.

The main advantage of this method is that the equipment is portable, therefore on- site fabrication can easily be accomplished. In addition, low cost molds can be used and complex shapes with reverse curves can be formed more easily than the hand lay-up process.

The spray-up process is used in making boats, display signs, truck roofs and other roofings, and in lining tanks. Large objects with connected parts can also be produced by this technique.

FILAMENT WINDING

In this processing technique continuous strands of fiber are used to achieve the maximum mechanical strength. Rovings or single strands of glass are fed through a resin bath and wound on to a rotating mandrel. The resin must have low viscosity such as 1 to 10 Pa.s to impregnate the fiber bundle easily. Preimpregnated rovings can also be used in the process. Some tension must be applied to the fiber to minimize voids and to have compact winding. When an external force is applied to a cured product, it is more or less equally shared by each strand wound under constant tension. This naturally imparts a superior strength to filament-wound products.

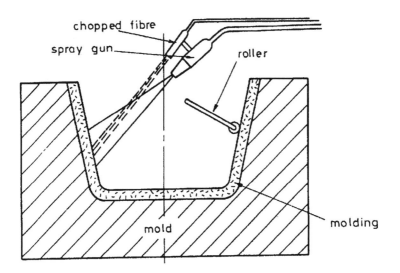

Figure 7. Spray-up technique [from Ref.28].

Fibers can be arranged to traverse the mandrel at different angles with different winding frequencies in a programmed manner accomplished by automatic control or computer aided control.

Cylindrical objects can be easily manufactured by this process as shown in Figure 8. However, items which have some degree of symmetry about a central axis such as conical shapes, isotensoids, cups, multi-sided box-like structures, and even spheres can be produced by this technique.

COLD MOLDING

In this technique resin is poured onto a mold and a small pressure is applied. Curing takes place at room temperature.

Cold Press Molding

Mold release agents and gel-coat are applied to the mold surface and then a glass mat is laid on. Then the resin containing initiator and accelerator is poured on the mat as shown in Figure 9. Several layers, if needed, can be built up this way. Then the mold is closed and a low pressure of about 150 kPa is applied to spread the resin over the mat.

Resin Injection

The mold surface is prepared and the fabric is laid down as in cold press molding and the upper part of the mold is clamped. Then the resin is injected under pressure into the mold as shown in Figure 10.

The catalysed resin and the accelerated resin are kept in separate tanks, they can be mixed just before injection, or mixed in the injection gun. The applied pressure is around 450 kPa. This process has the advantage of producing objects free of air bubbles and damaged fibers. The air is removed by letting some resin flow out from the mold cavity.

PULTRUSION

This is a process to produce composite materials in the form of continuous, constant cross-sectional profiles and more than 90 % of all pultruded products are fiberglass reinforced polyesters. Glass fiber ravings are drawn from an impregnation tank and then through a die of desired geometry as shown in Figure 11. The shaped ravings then pass through a tunnel oven for curing, and then pultruded composite is cut to proper lengths for storage.

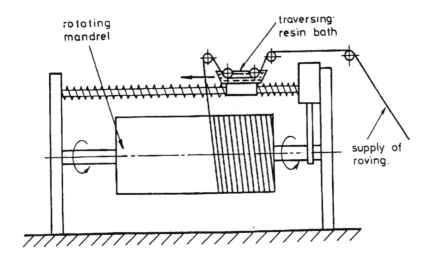

Figure 8. Filament winding [from Ref.28].

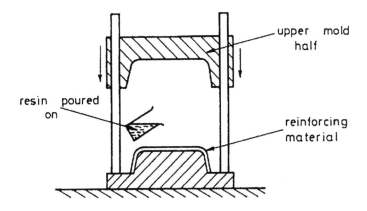

Figure 9. Cold press molding method [from Ref.28].

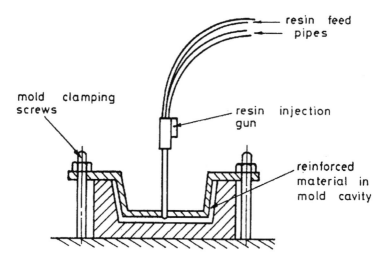

Figure 10. Resin injection process [from Ref.28].

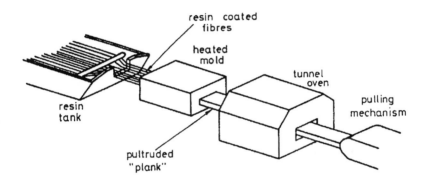

Figure 11. Pultrusion process [from Ref.28].

The resin used must have low viscosity around 2 Pa.s for fast wetting of the ravings, proper catalysts for a strong gel, and internal mold release agents. Typical line speeds range from 0.6 to 1.5 in/min. The mold temperature is between 375 to 400°K. A thermocouple control is needed at the dies to control the peak isotherm. Radiofrequency heating of material can also be used; the exotherm control is still necessary. Capital costs are higher than most processes, but labor costs can be minimal for long production runs.

Very thin and quite thick profiles of varying geometries can be manufactured by this technique. Some items made from aluminium or polyvinyl chloride are now substituted by pultruded composites. The products find applications in industry, transport, building materials including roofing, e.g., awnings, canopies, domes, and sheeting, and also in electrical and sporting-goods fields.

TESTS

Tests can be applied to analyze the physical properties of (i) uncured and cured resin, (ii) fiber, and (iii) composites.

Tests Related to Resin

The tests which give information about the resin are the followings;

Acid number	Molecular weight of the prepolymer is determined.
Viscosity	Flowability of the resin is found.
Infrared	Spectroscopy. Functional groups of the molecules making up the resin are determined.
High Pressure Liquid Chromatography	Information about the molecular weight distribution of the prepolymer is obtained.
Moisture	Excess moisture is deleterious to curing, so it must be kept below a certain level. Karl Fisher or any acceptable method can be used to determine it.
Gelation Conditions	Gelation time is determined by different methods to know about the rate of gelation

Peak Exotherm	The maximum temperature reached due to exothermic gelation reaction must be known to carry out a nonviolent curing.
Mechanical Tests	All fundamental strength tests such as tensile (ASTM D 651), compressive (ASTM D 695), shear (ASTM D 2344), flexural (ASTM D 90), impact (ASTM D 3998), and fatigue (ASTM D 256) can be applied to find out the mechanical properties of cast polyesters.

Tests Related to Fibers

The tests associated with fiber characterization basically are density (ASTM D 792, D 3800, D 1505), weight of unit length, filament diameter, thermal expansion, electrical conductivity, and the tensile strength (ASTM D 3379). In the case when a fabric is used, the pick count which is the number of tows per millimeter need to be established. In addition the weight per unit area, and the tensile strength (ASTM D 579) must also be known.

Tests Related to Composites

The composite product depends on moisture changes and thermal cycles. Moisture and temperature changes may cause delineation of fibers from the resin matrix. Thermal cycle tests can be planned depending on the type of resin and the thermal conditions where the product will be used.

Mechanical tests depend on the geometry of the product and the type of reinforcement. The recommended standards for the specified tests are given below [31].

Tensile Test	Unidirectional specimen (ASTM D 3039).
Compression Test	Unidirectional specimen or nonunidirectional laminates (ASTM D 3410), chopped fiber reinforced specimens (ASTM D 695).
Flexural Test	For nonreinforced and chopped fiber reinforced specimens (ASTM D 790).
Shear Test	For nonreinforced and reinforced specimens (ASTM D 2344).

MECHANICAL STRENGTHS

The mechanical and physical properties of UP castings and composites are given in Tables 2 through 7.

Table 2. Clear casting mechanical properties [from Ref. 32].

Material	Barcol strenth	Tensile strenth Mpa	Tensile modulus kPa	Elongation %	Flexural strenth Mpa	Flexural modulus kPa	Compressive strength MPa
Orthophtalic	-	55	3.45	2.1	80	3.45	-
Isophtalic	40	75	3.38	3.3	130	3.59	120
BPA* fumarate	34	40	2.83	1.4	110	3.38	100
Clorendic	40	20	3.38	-	120	3.93	100
Vinyl ester	35	80	3.59	4.0	140	3.72	-

* Bisphenol A

Table 3. Mechanical properties of fiberglass-polyester resin composites [from Ref. 32].

Material	Glass content wt %	Barcol strenth	Tensile strenth Mpa	Tensile modulus kPa	Elongation %	Flexural strenth Mpa	Flexural modulus kPa	Compressive strength MPa	Izod impact J/mm
Ortho-phtalic	40	-	150	5.5	1.7	220	6.9	-	-
Isophtalic	40	45	190	11.7	2.0	240	7.6	210	0.57
BPA fumarate	40	40	120	11.01.2	160	9.0	180	0.64	
Clorendic	40	40	140	9.7	1.4	190	9.7	120	0.37
Vinyl ester	40	-	160	11.0	-	220	9.0	210	-

Table 4. Effect of glass content on mechanical properties [from Ref. 32].

Material	Glass content wt %	Flexural strenth Mpa	Flexural modulus kPa	Tensile strenth Mpa	Tensile modulus kPa	Compressive strength Mpa
Orthophtalic	30	170	5.5	140	4.8	-
	40	220	6.90	150	5.5	-
Isophtalic	30	190	5.5	150	8.27	-
	40	240	7.58	190	11.7	210
BPA fumarate	25	120	5.1	80	7.58	170
	35	150	8.27	100	10.3	170
	40	160	8.96	120	11.0	180
Clorendic	24	120	5.9	80	7.58	140
	34	160	6.89	120	9.65	120
	40	190	9.65	140	9.65	120
Vinyl ester	25	110	5.4	86.2	6.96	180
	35	260	9.52	153.4	10.8	230
	40	220	8.89	160	11.0	210

Table 5. The effect of glass type and amount on mechanical properties [from Ref.32]

Type of glass fiber reinforcement	Glass content wt %	Density g/cm^3	Tensile strenth Mpa	Tensile modulus kPa	Elongation %	Flexural strenth Mpa	Flexural modulus kPa	Compressive strength Mpa
Neat cured resin	0	1.22	59	5.40	2.0	88	3.90	156
Chopped strand mat	30	1.50	117	10.80	3.5	197	9.784	147
Chopped strand mat	50	1.70	288	16.70	3.5	197	14.49	160
Roving fabric	60	1.76	314	19.50	3.6	317	15.00	192
Wowen glass fabric	70	1.88	331	25.86	3.4	403	17.38	280
Unidirectional roving fabric	70	1.96	611	32.54	2.8	403	29.44	216

FUTURE TRENDS

There is continuing work to combine unsaturated polyesters with other polymers in the form of interpenetrating networks or hybrid structures [33-37]. These attempts seem to open new fields of using unsaturated polyesters in the industry to produce items with improved physical and mechanical properties. Modified and new manufacturing processes are also expected to handle these new materials.

Table 6. Electrical properties of isophthalic polyester* 3.2 mm laminates with various fillers [from Ref. 32].

Material	Dielectric strenth short time V/mm	Volume resistivity 10^{13} Ω-m	Dielectric constant 1 MHz	Dissipation factor 1 MHz	Dielectric constant 60 Hz	Dissipation factor 60 Hz	Arc resistance			Track resistance V	Dielectric breakdown short time kV	Dielectric breakdown step-by-step kV
							Avg	Max	Min			
Calcium carbonate	15.0	7.8	4.10	0.007	4.19	0.0003	157	181	140	840	58	61
Gipsum CaSO₄	14.4	2.1	3.69	0.011	4.19	0.027	153	184	141	840	70	55
Aluminia trihydrate	15.4	2.6	3.67	0.009	3.89	0.011	183.5	184	183	860	67	51
Clay	14.4	6.4	4.08	0.018	5.10	0.057	182.5	183	182	840	59	57

* Vinyl toluene monomer

Table 7. Electrical properties of BPA fumarate polyester* 3.2 mm laminates with various fillers [from Ref. 32].

Material	Dielectric strength short time V/mm	Volume resistivity 10^{13} Ω-m	Dielectric constant 1 MHz	Dissipation factor 1 MHz	Dielectric constant 60 Hz	Dissipation factor 60 Hz	Arc resistance			Track resistance V	Dielectric breakdown short time kV	Dielectric breakdown step-by-step kV
							Avg	Max	Min			
Calcium carbonate	6.1	1.6	3.94	0.005	4.03	0.004	140	143	133	840	58	52
Gipsum CaSO₄	5.9	3.3	3.72	0.009	4.24	0.029	144	151	137	820	50	40
Aluminia trihydrate	11.8	3.3	3.64	0.008	3.93	0.025	182	184	181	820	55	52
Clay	12.6	3.5	4.08	0.023	5.11	0.053	183	184	181	840	61	43

* Vinyl toluene monomer

REFERENCES

I. C. Ellis, U. S. Pat., 1, 897, 977 (1933).

2. H. Dykstra, U. S. Pat,, 1, 945, 307 (1934).

3 T. F. Bradley, E. L. Kropa, and W. B. Johston, Ind. Eng. Chem., 29 (1937)1270.

4. G.Gündüz, Unsatutated Polyesters, to be published in "The Polymeric Materials Encyclopedia", ed: J. C. Salamone, CRC Press, Inc.

5. J. Kaska and F. Lesek, Prog. Qrg. Coat,, 19 (1991) 283.

6. H. V. Boenig, Unsaturated Polyesters: "Structure and Properties", Elsevier Pub. Co.,1964.

7. G.Gündüz and A. Deniz, Polym.-Plast. Technol. Eng., 31(1992) 221.

8. G.Gündüz and Ş. Öztürk, Polyal.-Plast. Technol. Eng., 33 (1994) 245.

9. J. L. Yu, Y. M. Liu, and B. Z. Jang, Polym. Composites, 15 (1994) 488.

10. J. F. Yang and T. L. Yu, J. M. S. - Pure Appl. Chem., A31(1994) 427.

11. J. Frados, "Plastics Engineering Handbook", Van Nostrand Reinhold Co., Fourth Ed.n, 1976, p. 444.

12. G.Gündüz, Steel-Fiber Reinforced Polymer Impregnated Concrete, in "Handbook of Ceramics and Composites, Vol.: 2 Mechanical Properties and Specialty Applications", ed.: N. P. Cheremisinoff, Marcel Dekker Inc., 1992, Chapter 6.

13. J. Agranoff, "Modern Plastics Encyclopedia 1976-1977", Mc-Graw Hill Book Co., pp. 138-222.

14. A. de Dani and H. V. Blake, "Glass Fibre Reinforced Plastics", George Newnes Ltd., 1963, p. 98.

15. E. N. Doyle, "The Development and Use of Polyester Products", Mc-Graw Hill Book Co., 1969, p. 309.

16. ibid..pp.312-313.

17. C. B. Bucknall, 1. K. Partridge, and M. J. Phillips, Polymer, 32 (1991) 786.

18. E. J. Battkus and C. H. Ktoekel, Low Shrink Reinforced Polyester Systems, in Polyblends and Composites, ed: P. F. Bruins, Applied Polymer Symp., No.15, 1970.

19. V. A. Pattison, R. R. Hindersinn, and W. T. Schwartz, J. App. Paly. Sci., 19 (1975) 3045.

20. L. Suspene, D. Fourquier, and Y. S. Young, Polymer, 32 (1991)1593,

21. Engineered Materials Handbook, Vol. 1, Composites, ASM International, 1987, p. 158.

22. V. 1. Szmercsanyi, L. K. Marcs, and A. A. Zahran, J. App. Poly. Sci., 10 (1966) 513.

23. T. L. Yu and S. C. Ma, J. Macromol. Sci.-Pure App. Chem., A30 (1993) 293.

24. Modern Plastics Encyclopedia, Mc-Graw Hill Pub. Co., 1970-1971, p. 196.

25. R. Gächter and H. Müller, "Plastics Additives Handbook", Hanser Pub., 1987, p. 128.

26. H. N. Hasipoğlu, H. Galip, G.Gündüz, Ö.Özdemir, and D. Turhan, The First Turkish Chemical Engineering Congress, Tebliğ Kitabl, 1.

Cilt, 13-16 Eylül, 1994, ODTÜ, Ankara, p. 590.
27. Ref.15.p.308.
28. R. J. Crawford, "Plastic Engineering", Pergamon Press, 1981, Chapter 4.
29. Ref.21.pp.165-166.
30. Ref.21.pp.157.
31. Ref.21.pp.289-301.
32. Ref.21.pp.90-96.
33. J. H. Kim and S. C. Kim, Polym. Eng. Sci., 27 (1987)1252.
34. B. Das,T.Gangopadhyay,and S.Sinha,Eur.Polym.J.,30(1994)245.
35. M. S. Lin, R. J. Chang, T. Yang, and Y. F. Shih, J. App. Polym. Sci., 55 (1995)1607.
36. J. F. Yang and T. L. Yu, J. M. S. - Pure App. Chem., A31 (1994) 427.
37. J. L. Yu, Y. M. Liu, and B. Z. Jang, Polym. Composites, 15 (1994) 488.

2

Processing of PVC

Tülin Bilgiç
Petkim Petrochemicals Research Center; Turkey

PROPERTIES OF PVC

General Properties

Commercial PVC is generally produced by addition polymerization. It may be produced by a variety of techniques such as suspension, emulsion, micro suspension and bulk. PVC, as normally prepared, is a white granular material, ranging in particle size from 5-400 microns and with apparent bulk densities of 0.5-0.8 g/cc. Emulsion type PVC is usually smaller in particle size. This type of PVC is also called dispersion type. Dispersion type PVC is grounded to further decrease particle size and this type of PVC is called paste type. Paste PVC particles are irregular in shape and have relatively high surface area. Both dispersion and paste type PVC are suitable for plastisol applications.

Uncompounded PVC is tough, brittle and has relatively poor heat stability compared to other thermoplastic materials and thus is never used without some modification. Providing that the plasticizer level is low, most PVC compounds burn slowly and tend to be self-extinguishing. Its high flash ignition temperature is another advantage. The physical properties of the compounded material are a function of both the resin and the compounding conditions. Plasticized PVC is the one of the most versatile plastic materials available. Compounding gives a very wide range of applications to PVC which approaches to that of rubbers and engineering plastics. Fire retardant properties, weathering resistance, excellent clarity and good flexural strength of PVC are its good qualities. The limitations are that, it degrades at elevated temperatures, can be corrosive to processing equipments, has relatively higher density than other plastics and susceptible to solvents. Table 1 gives a general idea of PVC properties.

Table 1. The properties of rigid and plasticized PVC [from Ref. 1,2].

Property	Rigid PVC	Plasticized PVC [from Ref. 1]
Density range	1.3-1.4 g/cc	1.1-1.7 d/cc
Specific heat	0.25 cal/g $^\circ$ C	-
Sag temperature	78°C	-
Milling temperature	160°C	140-150°C
Coeff. of linear ther. exp.	$5x10^{-5 \circ}C*$	-
Dielectric constant 30oC 60 cps	3.7	5.85
Power factor 30oC 60 cps	1.25%	-
Heat distorsion at 30 kg	80°C	-
Thermal conductivity	$3.94x10^{-4}$ $^\circ$C	-
Vicat softening point	-	58°C
Shore Hardness (D)	-	47
Tear resistance	-	8500 kg/m

* ~72°C Medium mol.weight PVC, 26% Dioctyl phthalate, 2.2% mixed stabilizer.

Chemical Properties

Addition polymerized PVC has largely head-to-tail arrangement of vinyl chloride units with a helical structure of $C_{28}H_{42}Cl_{14}$ repeating units.

$$(-CH_2-CHCl-CH_2-CHCl-)_n$$

The highly electronegative nature of chlorine leads to rigidity of chains resulting in a tough polymer. On the other hand, chlorine atoms are sufficiently bulky to separate the polymer chains which lead to less cohesion and increased freedom of molecular movement during plastization. PVC molecules show a low degree of branching and the extent of branching ranges from 0.5-20 branches per 1000 carbon atoms. PVC is slightly crystalline, mainly syndiotactic, but with so low a degree of order that only small crystallites are formed. The crystalline material content is about 2-10%. Cristallinity is influenced by thermal treatment and can be increased by polymerization at low temperatures. Both branching and crystallization depend on polymerization temperature (3). The chlorine atom also accounts for the fire extinguishing properties of PVC but it adversely effects the thermal stability of the polymer. PVC is fundamentally unstable to heat and light and loses hydrogen chloride by an autocatalytic reaction. HCl formation during degradation plays a catalytic role in PVC thermal degradation and it also causes corrosion problems.

The PVC chain also contains fragments of initiators emulsifying or suspending agents or other polymerization recipe ingredients as end-groups. End-groups may also be formed as a result of terminating reactions.

PVC is not hygroscopic and therefore normally does not absorb moisture. Commercial PVC contains 0.2-2% volatiles. Water absorption at 25^0C in 24 hours is about 0.05 - 0.10%. At high temperatures such as 100°C water absorption may be $> 10\%$.

PVC is soluble in cyclohexanone, dimethyl formamide, nitrobenzene, tetra hydrofuran. PVC is resistant to sulfuric, nitric, hydrochloric acid, sodium hydroxide, sodium hypochloride. It is not recommended for aromatic and or chlorinated hydrocarbons such as mono ordichlorobenzene, ketones, and alcohols. PVC wishstands to boiling water up to 140^0 F, moderately resists to detergent water and is not recommended for greases or oils (4).

As PVC is not soluble in its monomer, it precipitates as polymerization proceeds and eventually microstructural PVC grains are formed (5-6). The precipitation and agglomeration characteristics of PVC consequently defines the porosity, particle shape, particle size and distribution of the unique PVC particle structure. The microstructure of PVC particles depends on polymerization conditions such as temperature, type and quantity of dispersing or emulsifying agent and agitation conditions (7). A general microstructure model is not yet available, but there have been some attempts for modelling particle growth and thus the micro structure (8-15).

PVC particles may be categorized simply in three groups. Low porosity spherical particles, medium-high porous irregular particles, porous particles covered with a less porous layer. The microstructure of PVC plays a vital role during processing. Plasticizer up take rate increases as particle porosity and shape irregularity increases. Reducing particle size also increases plasticizer up take rate. The fusion rate of non porous PVC particles are relatively higher than porous particles due to the low thermal conductivity of air inside the pores. Flow characteristics and bulk density of PVC is influenced by particle shape, size, distribution and porosity of particles. With spherical particles the bulk density increases as particle size and/or porosity decreases.

Molecular Weight

The average molecular weight of PVC is usually expressed as viscosity number or K value. Both expressions are calculated based on viscosity measurements. The solvent, concentration and temperature of test affects the results. The molecular weight of PVC ranges Mw 40.000 - 480,000 and Mn 20,000 - 92,0000 corresponding to K value (DIN 53726) range of approximately 45-83. Molecular weight distribution Mw/Mn of commercial PVC ranges between ~1.9 - 5.2. The molecular weight of PVC is mainly affected by the polymerization temperature. Reducing the temperature increases the molecular weight, chain transfer agents also be used to adjust the molecular weight.

Thermal Properties

Commercial addition type PVC is mostly amorphous therefore do not show a crystalline melting point. PVC generally softens at 75 - 90°C. The glass transition temperature of PVC is dependant on polymerization temperature consequently on molecular weight. For polymerization temperature range of 50 - 90°C, the glass transition temperature Tg is 85 - 80°C respectively. The degradation temperature of PVC depends on many parameters such as amount of small molecular weight portion, small particle size portion, amount of porous and non porous particles, impurities etc. PVC usually degrades at ~195°C temperatures. The usual service temperature of PVC is 65 - 80°C. The specific heat of PVC is approximately 0.25 cal /g°C and the thermal conductivity is 3.94×10^{-4}cm. The calorific value of PVC is 1.9×10^4 K J /kg (16).

Density

The density of PVC differs by production process. Typical commercial rigid PVC densities range between 1.3 - 1.41 g/cc, which is relatively high in comparison with other thermoplastic materials. The density of PVC is influenced by the degree of crystallization. Therefore, temperature of polymerization, and the thermal history of the PVC affects the density. The density increases with reduced polymerization temperature. Density difference of approximately 0.5 - 1.1% may be observed between quickly quenched and slowly crystallized PVC samples.

Mechanical Properties

The mechanical properties of PVC depends on the compound properties The mechanical properties of PVC are given in Table 2. Izod impact strength decreases tensile stress increases as temperature reduces.

Table 2. Mechanical properties of PVC.

	Rigid PVC	**Plasticized PVC**
Tensile strength psi	5000-9000	1500-3500
Elongation %	2-40	200-450
Modulus of Elas. psi	800.000	1000
Izod Impact strength ft-lb/in	0.4-20	Varies
Compressive strength psi	8000-13000	900-1700
Rockwell Hardness	M 70	-
Flexural yield strength psi	10.000-16.000	-

Polymer Handbook covers the physical properties in more detail (17).

PROCESSABILITY

Rigid PVC is generally suitable for extrusion, injection molding, vacuum forming, calendering and bottle blowing. Thick sections may be difficult to fabricate due to high apparent viscosity of rigid PVC. Depending on the type of rigid PVC, processing temperatures range between 140-185°C, and generally mold temperatures of 20–120°C may be used. Mold shrinkage of rigid PVC is about 0.6% (0.004 in/in) (18). Apparent viscosity normally reduces as shear rate increases (non Newtonian behavior). For compression molding, a pressure range between 55 -140 kg/cm^2 may be necessary.

Plasticized PVC compound may be injection molded, extruded, or calendered. In the plastisol form rotational casting, spreading, dipping and foamed applications are possible. The processing temperature range is 150 - 200°C approximately. Linear mold shrinkage ranges from 0.01 to 0.05 in/in depending on the type of PVC and compounding formulation. For compression molding, the pressure range is between 38 - 140 kg/cm^2.

COMPOUNDING

PVC compounding aims ease of processing, desired end product properties and low cost. A variety of additives may be used in the compounding of PVC. The interaction of additives with the polymer is a complex matter depending at least to some extent upon the structure of the PVC polymer itself and upon the way structure changes during the fusion process. Most PVC formulations are complex multicomponent systems and it is some times an art rather than a science to specify the right formulation.

Advanced statistical and experimental design techniques speed up the selection of the proper PVC compounding formulation (19, 20, 21).

PLASTICIZATION AND FUSION

Plasticization occurs as a result of the action of plasticizer on PVC. The physical properties of PVC gradually changes from a rigid solid to a soft gel or viscous liquid. The action of plasticizer on PVC is a complex phenomena and involves many interactions. Four principal theories have been proposed to account for the main effects plasticizers produce. These are lubricity, gel, mechanistic and free volume theories (22). For more detailed theories the reader is referred to several textbooks which cover the subject in more detail (23-25). Plasticization occurs in a number of steps. The steps of plasticization may not occur for every polymer-plasticizer combination but various steps of plasticization for two main PVC applications, namely plastisol and dry blending are shown in Figure 1.

The plasticizer wets and adsorbs to the surfaces (Figure 1a) then diffuses through the PVC pores and affects the contacted outer surfaces. The plasticizer acts on the points of attachment along the polymer chain and breaks the attachments which hold the polymer chains together, and consequently solvates the polymer at these points (Figure 1b). Depending on the degree of solvation, the polymer particles swell (Figure 1c). On prolonged storage or heating, structure breakdown occurs. The unattached plasticizer molecules facilitate the movement of molecules and the PVC particles swell and after the second order transition temperature (T_g) the boundaries of secondary particles break (Figure 1d), the diameter of PVC particles changes. The change of particle size with temperature is shown in Figure 2. At this stage a dynamic equilibrium between solvation-desolvation and continuous exchange of plasticizer molecule attracted to the polymer chain occurs, the resin molecules aggregate and disgregate. Plasticizer affects the amorphous regions or crystal imperfections. If sufficient solvent power exists some kind of molecular order may also be observed by disappearance-reappearance of the crystallites. The salvation action of plasticizer increases and the gel forms. Gel is a semisolid state with enough integrity to hold itself together. It exhibits pseudo-elastic properties. Gelation could mean different things in different contexts but here, a continuous semisolid polymeric matrix is meant (Figure 1e). On heating over K value of PVC$+100°$C temperatures the boundaries of PVC particles completely disappears and fusion occurs. The physical changes at the upper end of the gelation region and the onset of fusion is different for plastisols and dry blends. The plastisol thickens and its haziness disappear, the salvation action of plasticizer increases, it passes through a fluid point where it is completely clear then it turns from liquid to solid at a short interval of time defined as gel point. If heating is continued fusion occurs. Fusion is the stage where a homogeneous phase at molecular level between polymer and

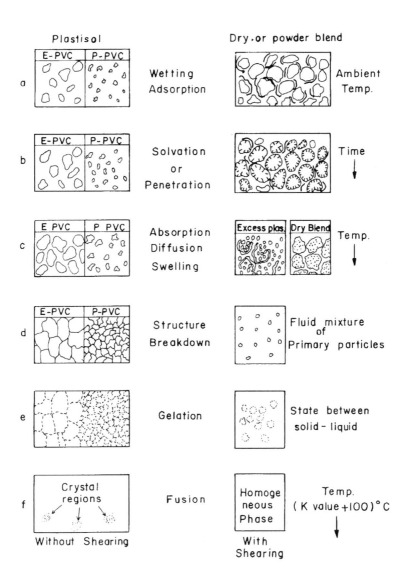

Figure 1. The changes observed during plasticization and fusion of PVC.

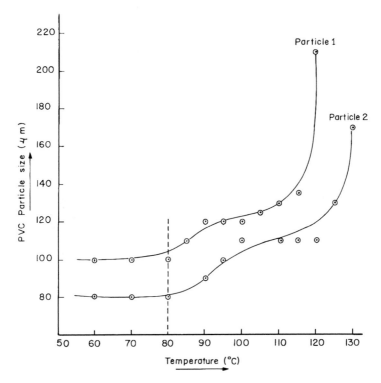

Observation made under optical microscopy transmitted light,
4 °C /min. heating rate. Rigid particles (with K value 65), in
excess amount of DOP.

Figure 2. The change of particle size with temperature
during plasticization.

plasticizer is reached. Beside heating, shearing forces are also necessary for molecular deformations and homogenization. Without shearing only the amorphous regions of the PVC is affected by the action of plasticizer therefore the crystal fragments are observed in the fused phase (Figure 1f). Particularly the fusion of dry blends requires shearing in addition to heating.

The gelling and fusion processes may be investigated by a couple of techniques and the change of plasticization steps are explained based on relative observations. The simplest method is the determination of viscosity changes of the system over the temperature range $25\text{-}200^{\circ}C$. When heat and shearing forces are applied to PVC, particle flow and molecular deformational flow take place. If the PVC dry powder blend is placed in a mixing chamber and subjected to heat and shear, the fusion stages may be observed by torque changes. A typical time-torque curve is shown in Figure 3. Usually two maxima in torque I and II are observed. Peak I occurs due to filling of the mixer. The maximum torque depends upon the applied filling conditions I. Maximum torque is reached after a few seconds and it indicates that the mixing chamber is filled with sample. As the mixing process continues, PVC particles are heated and sheared, and the viscosity decreases due to an increase of temperature and the first minimum is reached. After this point, the plasticizer swells the PVC particles, increases the particle diameter, causing the torque to increase. During the torque increase stage fusion begins and inter-particle boundaries disappear and the PVC particle inner agglomerates are distributed among the fused PVC phase. As the number of PVC particles fuse the torque increases and reaches the maximum II where the PVC mass is a homogeneous gel with completely entangled polymer chains. This is the point where fusion is completed and consequently the development of physical properties begins. If mixing is continued the stock temperature increases causing the torque to decrease to another minimum torque, which is an indication of mixing time for the PVC stock. After this minimum, molecular linkages and cross- links occur as a result of degradation, and hence torque increases considerably.

COMPOUNDING ADDITIVES

Plasticizers

Plasticizers are materials which increase the workability, flexibility or distensibility of PVC (26). Plasticizers may also lower second order transition (T_g) temperature or elastic modulus of the plastic. The PVC is unique in its acceptance of large amounts of plasticizers with gradual changes in physical properties from a rigid solid to a soft gel.

In general, plasticizers for PVC resins are esters of aliphatic and aromatic di and tricarboxcylic acids and organic phosphates. High molecular weight alkylaromatic hydrocarbons, chlorinated aliphatic hydrocarbons and

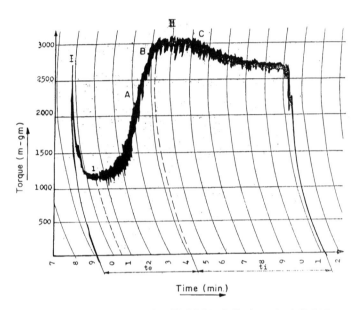

Brabender Plasticorder PL 2000-3 Fusion Head, 170°C,
Rigid PVC (K value 65), lubricants and mixed stabilizers.

Figure 3. Time-torque curve during PVC fusion test.

occasionally some other types are used as secondary plasticizers. The properties of some commonly used plasticizers are given in Table 3. A good plasticizer should have low volatility, light color, neutral reaction, resistance to hydrolysis, insolubility in water, flame resistance and non toxicity. While plasticizers must be chosen with regard to the properties desired in the final product, a further requisite for their use in plastisol is that the plasticizer must have little solvent action on the resin at room temperature, but good solvency at elevated temperatures. The lack of solvent action at room temperature is of primary importance in the preparation of plastisols which are to be stored prior to use. If solvent action takes place during storage it may cause an excessive increase in viscosity.

As no plasticizer can satisfy all these properties, several plasticizers are mixed to achieve the desired properties of the final product. The compatibility of plasticizer with PVC resin is also an important parameter in plasticizer selection (27).

There are many ways in which plasticizers are incorporated into PVC. Dry blending, hot compounding, plastisol and solvent casting are more frequently used techniques. Hot compounding is usually applied to rigid PVC types, which involves the mixing of plasticizer resin and other additives in an intensive mixer followed by heating and fluxing in two- roll mixers. At the end of this process a fused gel called "crepe" is obtained.

The type of plasticizer and its interaction with PVC is important in selecting the polymer/plasticizer ratio (28). The PVC properties which affect this interaction are size distribution structure, shape, porosity of particles molecular weight and distribution of PVC and other chemicals used in the compounding recipe. Plasticizers are usually added in amounts greater than 20 parts per hundred parts of resin (phr). The reverse effect, unplasticization, is observed at lower concentrations. For most resin-plasticizer combinations at room temperature, a plasticizer threshold concentration must be passed before the normal plasticizer has an effects on physical properties. The extent to which T_g is depressed depends on the amount of plasticizer present. The usual concentration range for plasticizers is 20-85 phr.

Lubricants

Lubricants are added to PVC compounds to facilitate processing and to permit the control of processing rate. They prevent sticking and may also function as antistatic agents.

The effectiveness of lubricants depends on their solubilities in the PVC compound. As the insolubility or incompatibility of lubricant increases they exudate to the surface during processing and act as an external lubricant. Not only the insolubility and incompatibility but also the excess amount of lubricant also functions as external lubricant reducing friction,

Table 3. General Characteristics of Commonly Used PVC Plasticizers.

Plasticizers (a)	Molecular Weight	Boiling Range (°C) (Pressure mmHg)	Melting Point (°C)	Refractive Index n_D (Temp. °C)	Relative Fusion Temperature (°C)	Note
Diisobutil adipate (DIBA)	426	349 (760)	-43 / -84[b]	1.451 (25)	133	Good low temp. properties. Poor volatility and extraction resistance.
Diisodecil adipate (DIDA)	258	135-147 (4)	-20	1.43 (25)	58	
Di-2-ethylhexyladipate (DOA)	371	214 (5)	-70[b]	1.447 (25)	111	Very good low temp. properties. Water sensitive than DOP. Non toxic.
Di-2-ethylhexylazelate (DOZ)	413	237 (5)	<-73	1.45 (25)	118	Better low temp. properties. Less water sensitive, volatile than adipates. More expensive.
Tributil-o-acetyl citrate (TBAC)	402	173 (1)	-80	1.441 (25)	102	
Tri-(2ethylhexil) acetile citrate	571	225 (1)	...	1.441 (25)	102	
Butylptthalyl butyl gicollate	336	219 (5)	...	1.489 (25)	75	
Polyethylene glycol di-2ethylhexonoate	Varies	219 (5)	-65	1.489-1.525 (25)	132	Low volatility Better low temp. properties.
Dimethyl-o-phthalate (DMP)	194	282 (760)	0	1.513 (25)	59	Rapid gelation. Easy processing.
Diethyl-o-phthalate (DEP)	222	296 (760)	-4	1.500 (25)	58	Very high volatility
Dibutyl-phthalate (DBP)	278	340 (760)	-40	1.492 (25)	62	Very high volatility
Butylbuzyl-o-phthalate (BBP)	312	240 (15)	...	1.54 (20)	69	
Di-2-ethylhexyl phthalate (DOP)	390	231 (5)	-46[b]	1.486 (20)	84	Standard primary Plasticizer.

Table 3. Continued:

Plasticizers (a)	Molecular Weight	Boiling Range (°C) (Pressure mmHg)	Melting Point (°C)	Refractive Index n_D (Temp. °C)	Relative Fusion Temperature (°C)	Note
Diisodecyl phthalate (DIDP)	446	255 (5)	-48	1.485 (20)	94	Heat resistent, better electrical properties, lower volatility and aquous extraction than DOP.
Di iso octyl phthalate (DIOP)	390	231 (4)		1.488 (20)	85	Standard primary plasticizer.
Di capryl phthalate (DCP)	390	227-234 (4)	-60	1.480 (25)	88	
Di-n-decyl phthalate (DnDp)	446	261 (5)	...	1.485 (20)	267	
Tri-2-ethylhexyl phosphate (TOF)	434	220 (5)	...	5.938	77	
Tricresyl phosphate (TCR)	368	265 (5)	-35	1.556 (25)	79	
Di-2ethylhexylsebacate (DOS)	426		<-60	1.451 (25)	125	Secondary plasticizer.

a) Plasticizer abbrevations ASTM D 1600, DIN E 7723, 1987.
b) Pour point.

adhesion and sticking in the polymer metal or polymer interfaces. Therefore control of balance of type and amount of lubricant for optimum performance is very important. Lubricants cannot be categorized as external or internal since it is likely that they act in both ways depending on their concentration and the processing conditions applied.

For plasticized PVC the most commonly employed lubricants are normal and dibasic salts of stearic acid. Myristic acids, paraffin and other waxes, low molecular weight polyethylene and certain esters such as ethyl palmitate could also be used.

Internal lubricants are chemicals which lubricate the flow of PVC molecules within the melt. They are rarely necessary in plasticized PVC compounds but they are essential for unplasticized compounds. Wax and wax derivatives, glyceryl esters of fatty acids, long chain alcohols such as stearyl, cetyl alcohols and long chain esters such as cetyl palmitate are most commonly used. Concentrations depends on the number of lubricants used to get the desired balance of behavior. Total concentration rarely exceed 4 phr. The normal concentration of internal/external lubricants is between 0.2-2.0 phr.

Impact Modifiers

These chemicals are usually acrylic polymers such as methacrylate-butadiene-styrene (MBS) acrylate-metacrylate, acrylonitrile-butadiene-styrene (ABS) which all introduce an incompatible rubbery phase usually offering a synergistic effect to PVC impact improvement. Therefore, low levels of modifier could generate required high impact resistance without changing the mechanical properties of the compound. Impact modifiers are selected according to the desired compound properties, for example optical properties such as light transmission, haze, yellowness index, low craze whitening or color reversal. For transparent applications, impact modifiers with refractive indices matching that of PVC are preferred. A refractive index difference of as little as 0.001 could cause light scattering and consequently, haze.

Impact modifiers are usually prepared by grafting methylmetacrylate and styrene on a styrene-butadiene rubber in an emulsion process (29,30). The preparation and properties of these polymers has been reviewed (31). Impact efficiency is achieved by controlling the rubber substrate particle size between 1000 and 3000 Å in a narrow distribution range. The solubility-incompability balance determines the performance of impact modifiers. Typical opaque compound impact modifiers show high efficiency. High-efficiency impact modifiers are used in PVC conduit, injection molding components and calendered opaque films and sheets. Various types of impact modifiers are available (32,33).

Semicompatible plasticizing polymers such as chlorinated polyethylene (CPE), ethylene-vinyl acetate (EVA), vinylchloride graft polymers and inorganics such as stearic acid-coated calcium carbonate are also used for impact modification of PVC.

Weatherable modifiers are used when retention of compound properties during prolonged external weathering is desired in applications such as PVC siding and window profiles. Modifiers based on butadien, such as ABS, MBS are normally avoided, instead saturated polymer systems such as, acrylic impact modifiers, butyl acrylate or 2 ethyl-hexyl acrylate graft phases are preferred. These modifiers do not have double bonds and therefore are more resistant to UV degradation but are not as efficient as butadien containing impact modifiers. Another weather resistant modifier is chlorinated polyethylene. MBS and ABS are available as free flowing powders therefore they are commonly incorporated into powder blend feedstocks. More rubbery materials such as chlorinated polyethylene requires a mastication process. The quantity of impact modifiers depend upon the application field. As a general rule the higher the molecular weight of PVC the lower the level of modifier needed (34). For unplasticized PVC applications 5-15 phr impact modifiers are used.

Processing Aids

Processing aids are chemicals which assist the melt processing of PVC. They are dissolved in PVC contrary to impact modifiers which present a separate phase to be effective. Their role is quite distinct from the role of lubricants. Processing aids do not reduce melt viscosity, they improve the elastic behavior of the melt. These additives are characterized by a high level of compatibility with PVC and much higher molecular weights than PVC, therefore these long chains tend to bind together the PVC melt structure thus increasing the extensibility of the melt. They can also produce faster plastication. The effect of processing aids on PVC is reviewed (35). Processing aids are usually copolymers of methylmetacrylate and other acrylates. Copolymers based on styrene are also used. Use levels can vary widely, depending on the particular conditions the range is 0.8 to 20 phr, but the most common range is 1-3 phr.

Stabilizers

Stabilizers are used to inhibit the degradation caused by heat or light. PVC is thermally unstable and degrades rapidly at normal processing conditions. It may also be degraded slowly by light. Therefore additives inhibiting these effects are necessary.

Irregularities in PVC chain causes instability. PVC chain consists mainly of a head-to-tail arrangement and head-to-head structures degrade at lower temperatures (36). The presence of internal double bonds (37), chloroallyl and carbonyl-chloroallyl groups, oxygen containing groups, labile chlorine atoms and configuration and conformation of polymer chain also affects degradation. Small molecular weight moiety, impurities particularly heavy metal ions and branching accelerates degradation. Thermal decomposition begins with unsaturated terminal groups and leads to the formation of conjugated systems, radical chain reactions, chain scissions and recombination.

Unstabilized PVC, if exposed to temperatures more than 100^0 C will turn yellow, then brown and eventually black. The rate of degradation depends on PVC properties, temperature and processing conditions. Initially the discoloration does not deteriorate its mechanical properties but after a certain time it decreases mechanical properties. The discoloration is caused by chemical reaction which starts with the elimination of HCl from PVC molecule. HCl plays a catalytic role in PVC thermal degradation. Once degradation reaches certain extent HCl extracts from the chain. This process is called dehydrochlorination. This is a characteristic feature of PVC degradation and as HCl is a corrosive gas, degradation of PVC in processing equipment causes severe corrosion problems. Thermo-oxidative decomposition is a chain radical process with degenerate branching. In this process, unsaturated terminal groups play a minor role.

Degradation of PVC is accelerated by thermal heating, oxygen attack, UV radiation and mechanical shearing which also accelerates the process (38). Depending on their protecting mechanism stabilizers may be categorized into two groups, namely thermal and UV.

Thermal Stabilizers

There are mainly two types of stabilizers. Primary stabilizers are chemicals which are capable of reacting with labile structures and accept HCl to inhibit the initiation of degradation. Secondary stabilizers inhibit or reduce the degradation of PVC by accepting thermo-oxidative groups. These stabilizers are effective in the propagation step. Stabilizers that react with oxygen or peroxide are called antioxidants.

PVC can be protected with a variety of stabilizer systems. The choice of stabilizer depends on the probable decomposition mechanisms involved. Primary and secondary stabilizers are generally used in combination. The primary stabilizers are mainly acid acceptors such as metallic soaps. Typical examples are, metallic soaps such as cadmium, barium, and zinc stearates which also act as lubricants. These chemicals simply react with HCl and are able to reduce further degradation. The mechanism of stabilization and properties of various metal soaps are reviewed (39,40). Calcium and zinc stabilizers are low in toxicity and they are often prepared in the form of a dispersion in epoxidized oil. Barium and cadmium stabilizers are mainly solid products those which are liquid contain phosphite groups. The compatibility of PVC stabilizers containing heavy metals (41) and synergetic effects of metal stabilizers (42) are reviewed.

Organotin compounds such as sulphur containing thio-tin compounds, octyl-tin compounds also act as partial plasticizers and improve melt flow properties. The mechanism of stabilization is more complicated. There are three main groups of organotin stabilizers. Carboxylate derivatives, sulphur containing organotins, carboxylate-mercaptides most organo-tin stabilizers could be used in transparent applications. Octyl-tin compounds are regarded as non toxic. The amount of tin and ligand structure affects the stabilizing efficiency (43).

Tribasic lead sulphate is the stabilizer which also provides opacity and gives white pigmentation. It is a cheap popular stabilizer but presents toxicity problems. It does not alter the melt rheology but it is an effective heat stabilizer and the opacity it produces assist in protection against UV degradation. Other types of stabilizers which mainly fullfill antioxidant effect are also used including phosphites, nitrogen containing stabilizers such as N, N'-diphenyl thiourea, β-diketons, polyols, phenol derivatives. Stabilizers are added in the concentration range of <1 to 3 phr.

The stability of PVC could also be improved by copolymerization or polymerizing in the presence of stabilizer (44).

Ultraviolet (UV) Stabilizers

UV stabilizers are chemicals which reduce or inhibit degradation of polymers resulting from UV radiation. The absorption of UV radiation by PVC excites PVC molecules if the molecules stay at this state for sufficiently long, the photolytic scission of certain C-Cl bonds occur and eventually cause breakage of chemical bonds which consequently degrades the polymer. The PVC compounds intended for outdoor applications require UV stabilizers. The function of UV stabilizers could be described by simply absorbing the UV radiation and re-emitting it as heat without deterioration. The basic chemical groups of UV stabilizers used for PVC are benzophenones, benzotriazoles, salicylates, acrylonitriles, hindered amines and pigments. Hindered amine light-stabilizers are the most commonly used UV stabilizers (~40% of light stabilizer market) (45).

Benzophenones UV stabilizers, which are effective in thin films import little color and have low toxicity. These type of UV stabilizers, for example hydroxybenzophenones form quinoid structure upon absorption of light. The quinoid reverts back to hydroxybenzophenone with loss of energy as heat. Substituent groups affect the UV absorption and volatility of UV stabilizers. The benzophenones are comparatively low priced UV stabilizers but discolor during processing and weathering.

Benzotriazoles show higher absorption than benzophenones at equivalent concentrations. These stabilizers also form quinoid structures. They have lighter color than benzophenones but more expensive. The properties of these type of UV stabilizers in film application has been reviewed (46). Although they are not used much, salicylates, benzoates, oxanilides and various pigments such as carbonblack and titaniumdioxide are used for specialty PVC compounds. Carbon black functions as a UV absorber and titanium dioxide as a light-scatterer and opacifier.

The UV stabilizers are very efficient at low concentrations. They are commonly used in 0.2-0.3 phr range.

Fillers

Fillers are added to PVC compounds to mainly reduce the cost. Additional technical improvements are also desirable such as better opacity, electrical properties, UV resistance, thermal properties, reduced plate out and sticking. Fillers are wet by the resin molecules and secondary bonds form between them. The normal effect of filler is to stiffen the resin systems. i.e., increase mechanical properties. Small amounts of filler may behave abnormally, yielding soft products. After a certain amount of filler is added, the normal effect of filler is seen i.e., increase of modulus and hardness. Fillers also increase the melt viscosity with a resultant reduction in processing rates. Plasticizer levels are adjusted to compensate this problem.

The most commonly employed fillers in plasticized PVC are the precipitated forms of calcium carbonate ($CaCO_3$) and related minerals such as dolomite (calcium magnesium carbonate). $CaCO_3$ fillers are throated with fatty acids to improve physical properties of PVC, particularly flex life and elongation at break. The effect of calcium carbonate fillers on fusion properties of PVC is reviewed (47). Other fillers include china clay, calcined clay, asbestos, barytas, talc, alumina and kieselguhr and silicates. Micaceous-talc fillers in PVC which are polyblended with acrylonitrile rubber shows an extraordinary effect. Small amounts improve elongation and energy to break.

The fillers could be used up to about 60 phr and higher concentrations are sometimes used where the application permits. Asbestos and whiting may sometimes be used more than 100 phr. Evaluation of the effect of various fillers on the physical properties of the compound are complicated and depends on the other ingredients of the compounds (48).

Computerized techniques are sometimes employed for cost/performance evaluations. The particle shape, size and distribution, oil absorption, dispersibilty and impurities of fillers affect its performance. Use of fillers with fine particle size and a refractive index closer to PVC results in the least reduction of clarity. Surface gloss decreases with coarse types and higher loadings cause whitening. Surface treated fillers are more easily dispersed. The presence of heavy metal impurities such as iron or zinc are avoided since they cause thermal degradation of PVC.

Increasing Property	Fine	Coarse
Oil Absorption	↗	↙
Gloss	↗	↙
Whitening	↗	↙
Tensile strength	↗	↙
Elongation at break	↗	↙

Figure 4. The effect of filler particle size on PVC properties.

Colorants

Colorants are added to PVC formulations generally to improve customer appeal and to hide defects. Vinyls are colored internally through the incorporation of pigments or dyes or externally through the application of coatings.

Dyes are organic substances which are soluble in the compound. The major disadvantages are their tendency to migrate and ease of extraction particularly with plasticizers. Therefore, dyes find limited use in vinyls except at very low concentrations.

Pigments are fine particle powders which are not soluble in the compound. Pigments serve the several functions, namely decoration, stabilization and loading. Although the decorative feature is usually emphasized, the other functions are also important. Pigment usually consist of agglomerated particles, therefore, for proper and efficient coloration, these agglomerates have to be broken down and the primary particles of the pigment must to be distributed as uniformly as possible. Pigment dispersion techniques are reviewed (49). Generally, vinyl compounds are colored homogeneously without much trouble but occasionally occurrences of pigment separation either as "plate-out" or in streaks could occur. For better distribution of pigments, concentrated pigment masterbatches are prepared where pigment agglomerates are adequately distributed through high shear mixing processes. Masterbatches contain high concentrations of pigments possibly up to as much as 50%. Depending on the nature of agglomeration, high speed mixers developing a fair amount of frictional work, or roll-mixer Banbury and twin screw type extruders developing high rate of shear and kneading are used. Service requirement, processing conditions and cost are to be considered in selecting a colorant. Nowadays, post consumer recyclate properties are also influencing pigment selection. Some pigments may reduce the heat stability of PVC compounds, for example, pigments containing ions such as zinc, iron, manganese, cobalt, copper, chlorine and sulfur decrease heat stability whereas pigments containing lead, tin, barium, cadmium and calcium improve heat stability. Cadmium reds-yellows and phthalocyanine blues and greens have the best resistance to long-term discoloration.

Pigments are divided into two major categories as follows:

Inorganic Pigments

This group includes titanium dioxide (TiO_2), chromium oxide, ultramarine blue, molybdate orange, metallic powders and flakes. Inorganic pigments usually have better heat resistance, high opacity and low cost in comparison with organic pigments. As they have limited solubility they generally resist bleeding, migration and extraction. Inorganic pigments are generally superior in light stability and weathering resistance. They have low plasticizer absorption and easily dispersed. Inorganic pigments are most suitable for opaque applications. Higher levels are required in comparison with organic pigments to obtain the same degree of coloration. The usual inorganic pigments for dark colors may contain iron or other metal ions that can adversely affect the stability of PVC. Some organic pigments, although more expensive, contain only traces of heavy metals. Because inorganic pigments reflect more IR radiation, less heat is absorbed by the plastic, whereas organic pigments have a higher tinting strength and hiding power over TiO_2.

Organic Pigments

This group includes phthalocyanines, quinacridones and benzidines. They are more powerful colorants with low specific gravity which tends to decrease the cost of using high cost organic pigments. Organic pigments have superior brightness brilliance and transparency. Especially when fine particle size types are used at low levels. Organic pigments have inferior heat stability and lower plasticizer absorption values. Organic pigments are used in low concentrations 0.02-0.03 phr. The comparison of general characteristics of inorganic and organic pigments are given in Table 4.

Table 4. Comparison of general characteristics of pigments

	Inorganic	**Organic**
Solubility in compound	Lower	Higher
Plasticizer Absorption	Lower	Higher
Light stability, wheathering resistance	Higher	Lower
Brightness and brilliance Transparency	Lower	Higher
Heat stability	Higher	Lower
Resistance to bleeding migration, extraction	Higher	Lower
Specific gravity	Higher	Lower
Dielectric properties	Lower	Higher
Cost	Low	High

Antimicrobials, Fungicides

Antimcrobials are added to plastics to protect them from microorganisms, which can cause the plastics appearance to degrade, cause the product to get brittle or lead to product failure. The two most commonly used antimicrobials are 10, 10'- oxybisphenoxarsine (OBPA) and 2-n-octyl-4-isothiazolin-3-one. Non-metallic type OBPA based antimicrobials are used in PVC compounding. Since plasticizers are susceptible to microorganisms, flexible vinyl compounding requires the usage of greater amount of antimicrobials.

As indoor air quality concern is growing, the usage of antimicrobials especially for carpeting, draperies, vinyl wall coverings and air conditioning gaskets are increasing. Antimicrobials that also control fungi are preferred. Protection from microbiological attack for applications in hot, humid climates are more essential. Zinc pyrithione based products function both as bactericide and fungicide. The most commonly used antimicrobials in PVC include organotins, mercaptans, quaternary ammonium compounds, arsenics and copper compounds. Arsenics, n- trichloromethylthio phthalimide are particularly effective against pink staining. Antimicrobials are used in a wide range depending on the plasticizer and other compound ingredients, the potency of bactericide, fungicide and the exposure conditions. Due to number of interactions involved, reproducible test results may not be obtained easily and evaluations based on relative observations are made. More reliable test methods are being investigated (50).

Antistatic Agents

Antistatic agents or antistats are chemicals which dissipate static electrical changes that accumulate on the surface of plastics. As plastics are insulating material they cannot independently dissipate an electrical changes build upon them.

Static electricity cause problems such as sparking, dust attraction and interference during processing. Spark discharge can be disastrous in environments containing flammable vapors. High level of electrostatic charges can also destroy information encoded in computer storage devices and interfere with electronic communication.

Electrostatic charges are most often generated on the surface of plastics through frictional contact with a second plastic surface, moving fluid (even air) or other surfaces that are in contact. Electrostatic charges are also produced or altered by generating new plastic surfaces such as milling, calendering, extrusion.

There are mainly two types of antistatic agents external and internal. External antistats are applied to the surface of the end product through techniques such as spraying, wiping or dipping and they are very effective is lowering surface resistivity but their action is not permanent. The additives used might migrate into the compound, rub off, abrade away or be contaminated with airborne dust.

Internal antistats are added to the compound and these additives are expected to migrate to the surface of the plastic to become effective. Internal antistatic agents have the ability to replenish the polymer's antistats protection through a process called blooming, the migration of the antistads to the polymer surface. Internal antistats could also interfere with other additives during mixing and processing but their effect is more permanent. Both antistats may decrease the coefficient of surface friction or transfer static charge to the ground or to the atmosphere.

The ionic chemical species resulting from the manufacturing process affects the antistatic properties of PVC. Thus, emulsion polymers have better antistatic properties in comparison with relatively clean suspension and more clean mass process resins. The volume resistivity of pure PVC is about 10^{11} - 10^{16} ohm-cm, whereas depending upon the application, lower values are desirable, therefore, antistatic properties are improved by compounding. Typical examples of some antistats used in PVC compounds areas follows:

Polyol Esters

These are the most common non-ionic antistats used in PVC, some examples are glycerol or sorbitan esters. They are generally nontoxic, biodegradable, FDA sanction, have relatively good heat and UV stability and modest cost. Polyolesters, for example glycerol monosterate are most commonly used, function as both internal and external lubricant. In rigid PVC applications there is usually enough external lubricants to decrease surface resistivity. For antistatic activity 0.1 to 0.5 phr levels are needed. For transparent applications, side chains without crystallization tendencies are preferred. In flexible PVC applications, if the polyol esters are compatible with the plasticizer, they are usually internal lubricants. Antistatic property could be strengthened through addition of 0.1 to 0.2 phr ethylene bis-steramide (51).

Ethoxylated Amines

Relatively long chain alkyl group ethoxylated amines are used to generate sufficient incompatibility and to cause blooming. Metal ions could be chelated to increase efficiency but they also reduce heat stability. For food contact applications maximum of 0.1% of ethoxylated amines may be used. Above this level decrease in heat stability occurs and cannot be used for FDA sanctioned areas. Amines could form unwanted color in combination with phenolic antioxidants such as BUT, BPA or some UV absorbers such as hydroxybenzophenone. Ethoxylated quaternary amine salts are more efficient they also function through hydrogen bonded surface network. The additives of this category must be investigated for heat stability effects.

Stearates

This is another category of lubricants that import antistatic properties as well as external lubricant and secondary stabilizer properties. Metals such as tin are ideal for forming hydrogen bonded networks on the surface and absorb atmospheric moisture. Alkyl sulfonates and phosphates are external lubricants. Generally their use posses less of a hazard to heat stability than amine salt of strong acids. Uses of barium and lead must also be checked as noted in the foregoing, for unexpected results.

Blowing Agents

Blowing agents are used to make foamed cellular PVC products. Blowing agents are added to PVC compounds for melt processing or as pastes. The blowing agents are gases which form as a result of mechanical, physical or chemical reactions and release at processing conditions to form cellular products.

Physical blowing agents are volatile liquids or compressed gases such as nitrogen, carbodioxide, air or low boiling liquids such as short-chain hydrocarbons. Short-chain hydrocarbons are dissolved in PVC compound and are volatilized because of changes they undergo during processing, by mechanical shearing, the release of pressure or heat of processing and give free gas to form cellular structures.

An inert gas such as CO_2 and a suitable plastisol could be mixed under pressure and on discharge the mixture becomes cellular. Micro-cellular products could also be prepared by using a filler which can be subsequently extracted by a solvent (52).

Chemical blowing agents are usually solids that decompose and generate gases by the heat of processing. Sometimes the gaseous chemical reaction product is used as blowing agent such as CO_2 generated from the reaction of water with an isocyanate which is present in the PVC compounding formulation.

The decomposition temperature and the amount of gas generated per gram of blowing agent is the most important characteristics of blowing agents. These properties determine the structure of formed cellular material. The decomposition temperature determines the cell structure, for example, an open cell structure is produced if decomposition occurs before gelation, and closed cell structures tend to form if the decomposition occurs after or at the some time as gelation. The distribution of blowing agent is also very important to form homogeneously sized cellular product. The blowing agents are selected according to the desired end product properties and processing conditions. In addition to cell structure, geometry, distribution and consequently the end product density, the toxicity of blowing agent is very important especially in closed-celled structure where the blowing agent exists for an elongated time, therefore nowadays nontoxic blowing agents are preferred such as inert gases. The impurities in blow agents is also very important as they might cause odor problems.

The concentration of blowing agent depends on the density required. The increase of blowing agent decreases the density of product but beyond certain level the cell structure becomes irregular and density cannot be reduced. If it is necessary to reduce density further, chemicals called "kickers" are used which lower the decomposition temperature ranges or accelerate the rate of decomposition. Kickers are mainly stabilizers such as tribasic lead phthalate, dibasic lead phthalate and dibasic lead phosphite which act as activators that reduce the temperature needed to produce the gas.

The chemical blowing agents may reduce the density of product from 0-5 g/cc to 0.16-0.3 g/cc by inclusion of some kickers. With proper selection of blowing agent and kicker system, densities down to 0.112 g/cc can be produced.

Flame Retardants

Flame retardants are used to treat a product so that its resistance to burning is improved or modified. PVC is inherently non-burning character because of high chloride content. It is a self extinguishing polymer. Some compound formulations are designed so as to add non-flammable properties. This is generally accomplished by addition of non flammable plasticizers such as phosphate esters or chlorinated hydrocarbons. Antimony oxide is widely used in PVC where low temperature properties of phosphates are aimed, levels of 1-3 phr antimony oxide are generally satisfactory. Sometimes 6 phr quantities are used. For transparent applications barium metaborate may be used.

PVC PROCESSING EQUIPMENT

A wide variety of different machines are used for the mixing and compounding of PVC compositions. General features of PVC mixing equipment are given below:

Mixing Machines

Tumble-Mixer: It is simply a drum which is then turned end over end.

Paddle-Mixer: The ingredients are agitated by means of blades, rotors in a vessel.

Ribbon Blenders: A type of blender which resembles a conveying screw.

Sigma Blade Mixers: These are comprised of a U shaped mixing chamber with Z shaped rotors.

Air Mixers: The agitation is accomplished by means of air fluidization.

High Speed Fluid Mixers: The agitation is accomplished by a rotor at the base of the mixing chamber, rotating at a speed up to few thousand revolutions per minute.

Ball Mills: These are used to grind filter or colorant agglomerates.

Colloid Mills: These mills impose intense frictional work and they are used for dispersion of agglomerated particles.

Mullers: These are comprised of a cylindrical shaped mixing chamber with one or two cylindrical rotors. Coarse particles may not be dispersed easily.

Roll Mills: These usually consist of two rolls, one in fixed bearing and the other in bearing which can be moved towards and away from the fixed one. Roll mills impose kneading action.

Internal Mixer (Banbury): This mixer originates from the rubber industry. The rotors are designed specifically for PVC compounding. During mixing intense kneading takes place.

Single-Screw Compounding Machines: The normal single-screw extruder is generally inadequate to homogenize PVC dry blend in a single pass but modified versions are used for this purpose.

Twin-Screw Compounding Machines: It is an extruder with two or more screws. It is possible to homogenize the PVC compound and extrude it at the same time.

Processing Equipment

Plasticized and unplasticized forms of PVC are processed with a wide variety of techniques. Comparisons of the general features of some PVC equipment are given in Table 5.

Table 5. General Features of PVC Processing Equipments.

Process	Type of operation	Complexity of shape	Width-Length Range (mm)	Thickness Range (mm)	Diameter Tolerances (mm)	Surface Finish	Tool Cost
Melt Forming							
Compression Molding	Intermittent operation	Simple Shallow	Only size of press	0.1-10	Good	Good (Edge flash)	Low
Laminating	Intermittent operation (slow)	Flat sheet only	Only size of press	0.4-1.5	Good	Good	Moderate
Calendering	Continuous Low output for thin foils	Flat sheet or foil only	0.5-2 wide any lenght	0.05-1.5	Good	Excellent	Very expensive
Extrusion	Continuous Low to high output depending on cross section tolerance	Constant section icluding hollow, re-entrant, complicated, multiple	0.1-1.5 wide and lenght	0.1->20	Moderate to excellent	Moderate to excellent	Low to moderate
Injection Molding	Semi continuous High output	Relatively small and compact, intricate re-entrant, threaded	1.5x1.5 10 gm-20 kg	0.1-10	Very good	Excellent	High
Post Forming							
Film Orientation (Uniaxial or Biaxial)	Continuous Moderate output	Flat or tubular film Also flat sheet intermittenly	0.-2 Any length	0.005-0.25	Good	Excellent	High
Pipe Sizing	Continuous Moderate to high output	Simple hollow Uniform Sections	5 mm- > 0.6m Any length	0.5-15	Good	Good	Low
Blow Molding	Semi-continuous Continuous Moderate to high output	Bottles Hollow, one end open relatively thin and uniform thickness	10 mm-0.6 m diameter ~400	0.2-10 lt	Good	External good Internal moderate	Moderate

Table 5. Continued:

Vacuum forming	Intermittent or Semi-continuous Moderate to high output	20 mm 0.1-10 mm diameter 2x2 m		0.1-10 Draw ratio ~ 5:1	Moderate	Moderate	Low
Pre-Forming							
Solvent Casting	Continuous	Thin sheets	Size of band and oven eg. 2 m wide	0.1-0.7	Good	Good	High
Rotational Casting	Intermittent Moderate to high output	Relatively thin hollows with opening also half hollows	1.5x2 ~5000 l	1-15	Moderate	Moderate	Low
Slush Molding	Intermittent	Small simple hollows of uniform thickness	Size of molds and oven	0.5-3	Bad	Bad	Low

REFERENCES

1. Garvin et al., Polymer Processes, Interscience Publ, NY, Chap XVI, 1956.
2. Bonotto, Walten, Modern Plastics, 40, (9), p.143, 1963.
3. Schildknecht, Vinyl and Related Polymers, John Wiley and sons, NY, 1952.
4. R. Greene, Corrosion Control in Proc. Ind., Me Craw Hill, NY, p.251, 1986.
5. P.V. Smallwood, Makromol. Chem., Makromol Symp. 29, p.1, 1989.
6. M.W. Allsoop, Pure Appl. Chem. 53, p.449, 1981.
7. N.Özkaya et al, Die Angew. Makromol. Chem., Vol. 211, p. 35-51,1993.
8. G.R. Johnson, J.Vinyl Technol. 2, p.138, 1981.
9. G.Weichert et al., Angew. Makromol. Chem., 164, p.59, 1988.
10. G.Weichert et al., Angew. Makromol. Chem., 165, p.35, 1988.
11. T.Y. Xie, A. E. Ham ielec, P. E.Wood, P. R. Woods, J.Vinyl Technol. 13, p.2, 1991.
12. P.V. Smallwood, Polymer 27, p.1609,1986.
13. M.J.Bunten, Encycl. Polym. Sci. Eng., 2nd ed., Vol. 17, p.295, 1985.
14. Butters, Particulate Nature of PVC, Appl. Sci.   Publ.   Ltd., Barking, p.17, 1982.
15. C. Kiparissides et all, J. Appl. Polym. Sci., Vol.54, No. 10, p. 1423-1438,1994.
16. VV. Schnabel, Polymer Degradation, p 55, Hanser Publ., Munchen, 1981.
17. Brandrup et al., Polymer Handbook, 3rd ed., John Wiley and Sons, N.Y., 1989.
18. A.Whelan, Injection Molding Materials, P.268, Appl. Sci. Publ. Ltd., N.J., 1982.
19. Brofman CM, J. Vinyl Tech., Vol. 10, No.3, P. 148,1988.
20. Carlson, Rubber and Plastics News, Vol.21, No.18, P.15-17,1992.
21. Brown, J.Vinyl Tech. Vol. 14, No.3, P. 161, 1992.
22. W.V. Titov, PVC Technology, Elsevier Applied Sci. Publ. 4th Ed, N.Y., 1986.
23. Sears and Darby, The Tech-of Plasticizers, John Wiley & Sons, N .Y., 1982.
24. Doolittleet al. J. Paly. Sci., Vol.2, P. 121-141, 1947.
25. Nass and Heiberger, Encycl. of PVC, Vol.1, Marcel Dekker Inc., N. Y. 1985.
26. ASTM D 883 "Definition of terms relating to platics", 1995.
27. Ramos L, Plast. Rubber Process. Appl. 130), P.151, 1990.
28. Howick, Plast. Rub, and Camp. Proces. and Appl., Vol.23, No.1, p.53, 1995.

29. US Patent 3, 793, 402,1974.
30. Kirk-Othmer, Encycl. of Chem. Tech., John Wiley & Sons, N.Y., p.390, 1981.
31. Purcell et al., Encycl. of Paly. Sci. and Tech., John Wiley & Sons, N.Y., 1976.
32. Ernest W-Flick, Plastics Additives, Noyes Publications, N.J., p.249-253,1986.
33. Lute J.T., Journal of Vinyl Technology, Vol.15, No.2, p.82-99,1993.
34. Bucknall, C.B., Tougheded Plastics, Applied Science, p.153, 1977.
35. Parker et al., Journal Vinyl Tech., Vol.15, No.2, p.62-68,1993.
36. N.Murayama, Y.Amagi, J.Polym. Sci., B-2, p.115, 1966.
37. Lucas R. et al, Journal of Applied Polymer Science, Vol.30.no.2, p.843, 1985.
38. Yano S, Journal of Applied Polymer Sci., Vol.21, No.10, p.2645-2660,1977.
39. Jerzy Wypych, PVC Stabilization, Elsevier, N.Y., p.41-75 1986.
40. Neiman, Aging and Stab, of Polymers, Consultants Bureau, N.Y., 1965.
41. W. Griebenow, Kunstst, Ger. Plast, Vol.82(l), 3-31, p.17, 1992.
42. Grossman, Jour. OfVinyl Tech., Vol.12, No.l, p.34-42,1994.
43. Kugele, Jour, of Vinyl Tech., Vol. 13, No. I, p.47-49,1991.
44. Gressier, Applied Makromol. Chem. and Physics, Vol.187, p.153-167,1991.
45. Ullmann's, Encyc. of Ind. Chem. Vol. A3, P.103, VCH Verlagsgesellschaft GmbH, 1985.
46. Inoue K, J. of Appl. Poly. Sci., Vol.50, No.ll, p.1857-1862, 1993.
47. Plasketl A.M., University of Technology, Loughborough (UK), The Effect of Calcium Carbonate Fillers on Fusion and Properties of Rigid PVC, Dissertation Abstracts 651 (6), Dec. 1990.
48. Mark's Encyclopedia of Polymer Science and Engineering, 2nd Ed. P.69, John Wiley and Sons, 1987.
49. Modern Plastics Encyclopedia, Me Graw Hill, 1991.
50. R.Borgmann Strahsen, Evaluating Micro Biological Susceptibility of Plasticized Films, Kunsts Toffe Plast Europe, Vol.84, No.2, p.158, 1994.
51. R.F. Grossman, Antistatic Agents, J. Vinyl Tech. Vol.15, No.3, P.164, 1993.
52. Matthews, Vinyl and Allied Polymers, London Iliffe Books, London, p.273-274, 1972.

3

Polyethersulfone (PES) and Its Processing

Kun Qi
Guangdong University of Technology, P.R.China

Rui Huang
Chengdu University of Science and Technology, P.R. China

INTRODUCTION

In the past few years, a whole range of resins and fiber-reinforced composites have been developed based on new, tough high-temperature thermoplastics. Polyethersulfone (PES) is one such polymer which is generating a great deal of interest, especially as a structural material occupying a unique place in the industry [1-9].

PES resins are thermoplastics having properties that place them high on the list of engineering plastics. Excellent electrical properties allow the added bonus of double insulation protection. They are highly resistant to most chemicals. Molded parts retain their shape at elevated temperatures, and have flame retardants properties. All these make PES plastic the best choice for appliance, electronic, communications, automotive, outdoor use, food-contact, and metal-replacement applications. In addition, they have an outstanding cost/performance balance [1-4].

PES may be reinforced with glass-fibers and carbon fibers to increase their tensile strength, stiffness, and dimensional stability [1,3].This chapter is limited to a discussion on un-reinforced PES with emphasis given to the processing of PES materials.

PES is also known as Victrex Polysulfone, which is produced at this time by ICI, Ultrason by BASF and Astrel by 3M, etc.[1,2,6]. Unless otherwise noted, all data discussed here are based upon PES resin supplied by Jilin University, Changchun, P. R. China. The structural repeat unit can be represented as <I>. Before processing, the PES resin was dried at 150°C for 4 hours to remove any absorbed moisture.

$$\langle\, I\,\rangle$$

PROCESSING PROPERTIES

Flow and Rheological Behavior

The use of PES material requires that it be self processed for injection and extrusion molding, etc. The knowledge of the rheological behavior of PES over an industrially relevant range of shear rate and temperature is essential for proper material processability, process design optimization and troubleshooting.

Shear Rheometry of PES Melt [10,11]

Flow curves - Melt rheological properties of PES were evaluated on a capillary instrument attached to a Shimadzu Universal Materials Testing Machine model AG-10TA. Viscosity curves measured at 315, 330 and 350°C and for shear rates ranging from 10 to 10000 1/s are presented in Figure 1. A typical pseudo-plastic behavior can be seen. That is, the melt viscosities of PES decrease with the increase of apparent shear rates.

The shear stress vs. rate plots of molten PES are shown in Figure 2. As expected, due to the strong non-Newtonian behavior, the variations of log (shear rate) with log(shear stress) are not linear. This suggests the power law index, n, will change with shear rate. The power law index values calculated are reported in Table 1.

The power law indices, n, in Table 1 are close to the value (n=0.53) reported by Saini [12]. As can be seen, the values often decrease with a decrease in temperature, and increase with shear rate. In other words, the lower the temperature and the higher the shear rates, the more non-Newtonian the behavior.

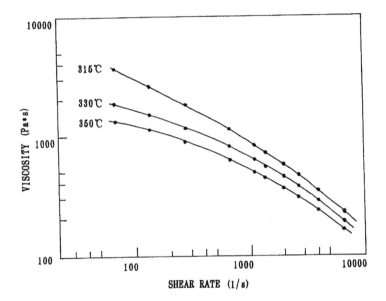

Figure 1. Viscosity vs. shear rate for PES.

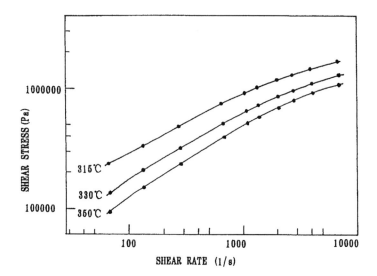

Figure 2. Shear stress vs. shear rate for PES.

Table 1. Flow index values of PES.

Temperature (°C)	Flow index values at shear rates of	
	10-999 1/s	1000-10000 1/s
315	0.50	0.33
330	0.58	0.41
350	0.63	0.40

Factors Affecting the Rheological Properties of PES

The temperature dependence of viscosity follows the Arrhenius equation. The flow activation energy (J/mol), E, for viscous flow can be measured from log(apparent viscosity) vs. 1/T plots, where T is the absolute temperature (K). The activation energy of viscous flow is found to decrease with increase in shear rate (Figure 3). But it can be seen that the viscosity of molten PES has a weak dependence on temperature, obviously due to the fairly lower viscous flow activation energy it has.

The viscosity of PES decreases with an increase in shear rate. PES as a polymer consists of semi-rigid molecular chains which have a strong affect on rheological behavior. Hence, a decrease in the degree of apparent viscosity of PES is less than that of a flexible-chain polymer, but higher than a rigid-chain polymer, with increase in shear rate.

Torque Rheological Property of PES

The torque-temperature-time diagram of PES, measured by a microcomputer-controlled Haake Torque Rheometer model SYS 40, is presented in Figure 4.

Torque Rheometer provides a flow curve that is a measure of the polymer's processability [9,13]. In Figure 4, the curves for TQ and T2 represent the relationship between torque and temperature, with time respectively. The points L, S and D were the times of loading, stabilization and degradation of PES, indicated automotive by Haake, during kneading, respectively. It can be seen from Figure 4, that the torque upon loading of PES is quite large, but stabilizes to a relative small value. Hence, the energy consumption of the molding machine during processing of would not be very large [14,15].

Both torque and temperature stabilize to a constant value in about 5 minutes (Figure 4). This suggests that in order to obtain a molten state, a molding machine with high length to diameter ratio, L/D, or a twin-screw

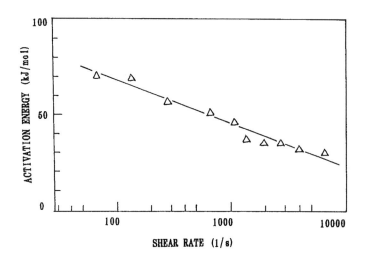

Figure 3. Variation of viscous flow activation energies with shear rate.

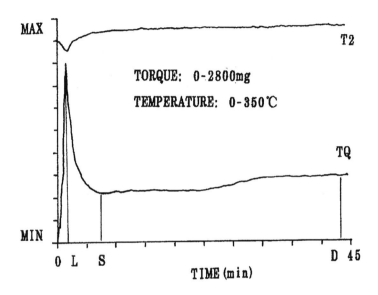

Figure 4. Torque (in meter-gramm) vs. the time at 335°C and 15 rpm.

and/or high pressure machine should be used in practical processing applications.

At a time of over 40 minutes (point D), a considerable increase of torque is observed. At this point, PES products turn black and the machine's energy consumption   during injection and extrusion molding increases, especially if the melt has a long retention time in the barrel [14]. When an increase in melt viscosity occurs with an increase in processing time, caused by prolonged shear, heat, oxidation and different forms of thermal degradation occur. This phenomenon is referred to as thickening [l6,17].

There is a relationship [l8] between torque (Nm), M, and rotor speed (rpm), S, which can be described by the familiar power law, and can be used for calculation of the power index, n. Figure 5 shows variation of equilibrium torque (torque at point S as in Figure 4) vs. rpm at 335°C. A power index obtained from Figure 5 is n=0.58, which is consistent with Saini [12] and capillary rheological results presented above.

Spiral Flow Tests

Flow tests are run in an actual injection molding machine. Results are usually direct, tangible, and easy to interpret. Typically the flow length mold forms a part in the form of a graduated spirals. The flow length is measured and can be related to flow conditions in an actual production mold. "Spiral Flow Length (SFL)" data is used especially for comparing various resins for ease of fill [13].

SFL data of PES were measured on a screw injection machine Model V7-4, manufactured by NB&C. The variation of SFL with injection rate, level 1 to 9, is illustrated in Figure 6. The SFL values are quite low at low injection rate levels. The solidification point of PES is very high, so lower flow rate through the nozzle resulted in fast solidification of molten PES in the spiral channel with a lower mold temperature. After a continuous increase of the SFL with injection rate, a equilibrium SFL is attained at an injection rate of about level 4. Injection rates higher than level 4 do not have an obvious affect on the SFL. The decrease in SFL at injection rate level 9 may be caused by leakage flow due to higher injection rates.

The increase in melt temperature causes an obviously linear increase in the SFL. A barrel temperature of 300-350°C is recommended in order to easy plastization and avoid the danger of degradation. Injection pressure compensate the pressure loss before and after the plastic enters the cavity, which gives specific flow rate and compacts the melt. Figure 7 shows the curves of injection pressure vs SFL. As shown, a suitable injection pressure is about 100 MPa.

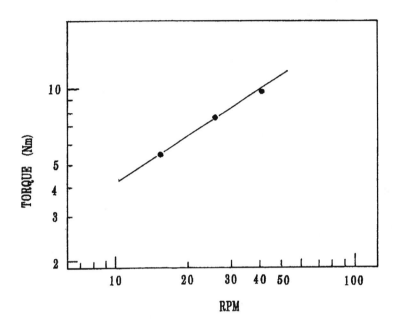

Figure 5. Variation of torque with rpm at 335°C.

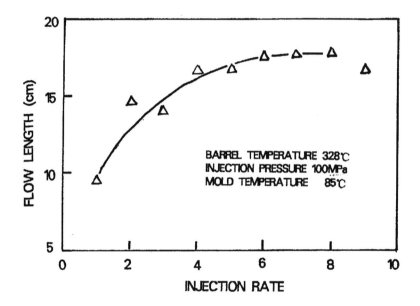

Figure 6. Variation of flow length with injection rate.

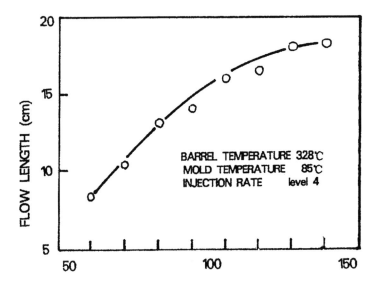

Figure 7. Effect of injection pressure on flow length of PES.

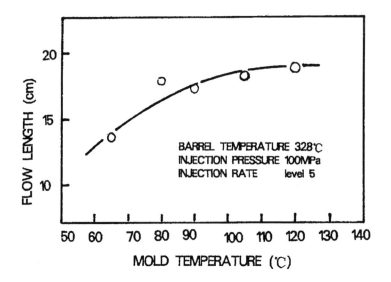

Figure 8. Mold temperature vs. flow length for PES.

The increase in mid temperature can cause a slight increase in the SFL. As the solidification point of MS is rather high, the mold temperature ranges from 60-130°C and does not have a notable affect on the rate of cooling (refer to Figure 8).

Effect of Rheological Behavior on Molding [11,15]

In plastics processing, adequate melt viscosities are required to make products using different processing methods. By adjusting the molding conditions, the processing flexibility can be achieved.

Due to its specialty in molecular chain structure, i.e., a semi-rigid-chain, the viscosity of PES is not very sensitive to changes in temperature or shear rate. Hence the flow performance cannot be increased effectively by shear rate or increases in barrel temperature and injection pressure. Since the SFL of PES is relatively low, it is difficult to mold parts with sophisticated and thin-wall structures.

Drying and Moisture Absorption

Pellet drying prior to injection molding and extrusion is necessary for many plastics. For a water absorbing plastic like PES, residual water could reduce the toughness of the molded part, and can result in streak line formation in the molded part, as well as degraded molecular weight during melt processing. In extrusion, strand breakage and frothing could occur due to the presence of residual moisture [14,15].

Usually the mass-transfer inside the polymer is the controlled resistance and the drying time is mainly controlled by the diffusion coefficient of the water in the polymer and the sample dimensions. The amount of water that can be removed depends on the humidity of the drying gas and the solubility of water in the pellet at the pellet temperature.

Key Parameters Affecting Pellet Drying [9]

Drying temperature and drying time have significant impacts on the drying process. First, the drying time is a strong function of the drying temperature. The drying time at a higher temperature is faster than at a lower temperature (14). Second, an increase in temperature decreases the equilibrium solubility as seen in Figure 9. This means that the PES pellet can be dried to low values with a higher temperature.

Pellet size also has an impact on the drying operation. Decreasing the length and the radius of strand-cut pellets increases the rate of drying somewhat. Compared to the effect of pellet temperature and time, the effect of pellet size is only secondary.

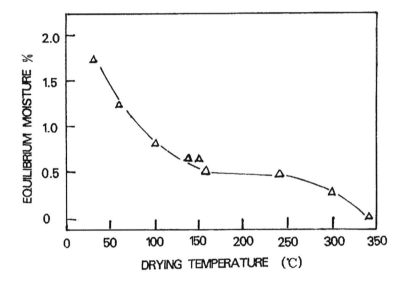

Figure 9. Effect of drying temperature on the equilibrium moisture level
in PES.

Figure 9 shows the equilibrium moisture in PES, dried with saturated air at different drying temperatures, measured by using a Computrac TMX water analyzer. The solubility of water in PES resin at room temperature was found to be 1.7%. The % equilibrium moisture goes down with an increase in drying temperature below 150°C, and then remains unchanged until 250°C. According to our data [14,15], pellet drying at 140-160 °C for 3 hr, prior to injection molding and extrusion is recommended.

Diffusion Coefficient [9]

For the diffusion experiment, the samples were prepared by saturating the pellets in two natural environments: 71% relative humidity at room temperature of 28°C and 80% relative humidity at a room temperature of 15°C for a period of time.

Certain assumptions [19-21] , including the use of Fick's diffusion law [22,23], are made. The diffusion coefficients were obtained by minimizing the root-mean-square error between the experimentally measured pellet moisture. Calculated values of the diffusion coefficient are 8.02E-9 cm^2/s and 7,79E-9 cm^2/s at 28°C and 15° C respectively.

The activation energy for the diffusion coefficient, obtained from an Arrhenius plot of the diffusion coefficient versus temperature, is 1700 J/mol in the 15°C to 28°C temperature range.

Thermal Stability

PES is one of the most important thermally stable aromatic polymers, for which the glass transition temperature, T_g, is so high that it can be used at high temperatures of around 200°C. Though it is significant to understand the features of degradation of the polymer in the solid phase because they are used practically as solid materials, it is much more important to know the degradation features of the molten polymer with respect to its molding processes. From this point of view, attention will be directed to the degradation process in the molten state.

Among the various methods used for studying thermal and thermo-oxidative degradation of PES, thermogravimetric analysis (TGA) and pyrolysis gas chromatography-mass spectroscopy (PGC-MS) have been used most frequently. These instruments enable comparisons of the relative thermal stability and thermal decomposition temperature, and give information about the degradation mechanism [9,24]. TGA also is an excellent technique for product characterization and quality control.

Data reported previous [9,24] were measured by using MAT445 type PGC-MS with a Curie point pyrolyzer and OV-1/C545 columns. No pyrolyzate was detected at temperatures below 358°C. The pyrolyzate at 590°C was composed mainly of sulphur dioxide and phenol, so it is believed that chain scission and hydrogen abstraction would be the possible degradation mechanisms of MS pyrolysis in an inert atmosphere.

Thermal Characterization: Decomposition Behavior vs Atmosphere

The results of the TG-DTG measurements [9,24], performed on a Perkin-Elmer DSC-7 thermal analyzer, show marked differences in the degradation of PES in a nitrogen versus an air environment. The sample heated in nitrogen leaved 20-30 % residue and in DTG curve there is one characteristic peak indicating a single mode of degradation; while in air atmosphere they burned off completely and there are two characteristic peaks indicating a double mode of degradation. Table 2 shows temperatures corresponding to characteristic peaks in DTG curves at the highest rates of degradation.

Table 2. Characteristic temperatures at the highest rate of degradation.

Temperature (°C)	Rate of heating, °C/min			
	10	20	40	80
in air (1)	596	620	644	651
in air (2)	589	593	606	617
in nitrogen	670	685	704	725

Effect of Heating Rate

In an air atmosphere, TG curves of PES were influenced by the heating rate still more than in a nitrogen environment [9,24]. PES has enough time reacting with oxygen when heated in air slowly to give regular TG-DTG curves. When PES is heated in air fast, the reaction can not coincide with the heating rate, and hence, a thermal lag develops. Temperatures at the beginning of weight loss gained from TG plots of PES in an air and in a nitrogen environment at various heating rates are given in Table 3.

Table 3. Temperatures at the onset of weight loss.

Temperature (°C)	Rate of heating, °C/min			
	10	20	40	80
in air	559	561	568	594
in nitrogen	565	595	603	613

Decomposition Kinetics

The activation energies [9] of degradation of PES are shown in Figure 10. In a nitrogen environment, activation energy remains a constant at about

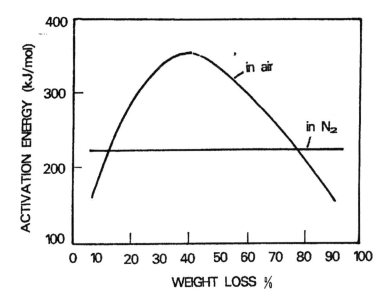

Figure 10. Curve of weight loss vs. activation energy for the
degradation of PES.

210 kJ/mol, with the progress in degradation. In an air atmosphere, thermal-oxidation degradation becomes a complicated situation. A maximum value for the activation energy of about 370kJ/mol appears at a 40% weight loss.

PROCESSING-STRUCTURE-PROPERTY RELATIONSHIPS

Effect of Processing on the Structure and Properties of PES

PES is an important engineering thermoplastic that displays excellent resistance to hydrolysis and oxidation and, at the same time, has excellent mechanical properties, good thermal stability, and toughness [1,25]. One of the major advantages claimed for PES is its ease of melt processing, which also implies that the material can be reworked and repaired several times by heat treatment. But as a high-temperature engineering thermoplastic, PES should, of course, be processed at very high temperatures, due to its high glass transition temperature (T_g=225°C). Thus, processing conditions may give rise to different thermal and thermo-oxidation processes that will affect the properties and structure of PES. These processes are even more important when PES is subjected to reprocessing, that is, to several processing cycles. Reprocessing is an activity that is especially common in the case of PES because of the high price of the raw material.

Because of the above, the structure and properties of PES can be greatly influenced by thermal processing cycles. The selection and control of the processing conditions is, therefore, essential in optimizing the structure and properties of finished parts. There are numerous studies on the stability of PES based on thermal analysis [25-27], as well as structure and property changes during the preheating process under different processes conditions (thermal-aged) [28,29]. However, only these studies conducted to investigate the charges in structure and properties of PES during real processing conditions are more important [16,17].

Kneading Processing [9,16,17]

Kneading of PES was carried out in an Haake Torque Rheometer Mixer at 335°C and at 15 rpm, 25 rpm, and 40 rpm during increasing periods of time to a maximum time of 1 hour. The results of kneading tests conducted at different times and different rotation speeds are summarized in Table 4. It is found that the color of PES samples that have been kneaded deepens. It seems that there are some changes occurring in PES, due to prolonged shear, heating and thermo-oxidation.

In a torque-temperature-time diagram, a slow but continues increase of the torque is observed. The torque also affected by the speed of the roller.

The molecular weight, MW, of PES samples subjected to different kneading process are also reported in Table 4. As shown, MW increases with time at the same speed of the roller (15 rpm), and it also increases with the speed of the roller after the same duration.

Calorimetric analyses were also performed on PES samples. The results indicate that the glass transition of all PES samples have no distinct change, and the glass transition point T_g remains constant irrespective of kneading conditions, as observed in Table 4. All these results indicate that the reaction is probably a grafting reaction, which takes place as a consequence of prolonged shear, heating, and thermo-oxidation during kneading, and further have any notable effect on the structure of PES.

Table 4. Results of kneading tests.

				Speed of Kneading Minimum Ultimate			
Samples No.	Temp.set (°C)	Roller (rpm)	Time (min)	Torque (mg)*	Torque (mg)*	T_g (°C)	Molecular weight
1	335	15	10	---	---	224.0	2.3E4
2	335	15	20	552	634	225.0	3.1E4
3	335	15	30	507	589	224.5	3.1E4
4	335	15	40	553	698	225.0	3.2E4
5	335	15	50	553	680	225.0	3.3E4
6	335	15	60	552	906	224.9	3.4E4
7	335	25	50	752	996	224.5	3.8E4
8	335	40	50	968	968	223.5	4.4E4

* Metric conversion 1000 mg=9.807 Nm

Successive Extrusion [9,17]

Successive extrusion was carried out in a extruding machine, with a screw diameter of 20 mm, at the extrusion conditions given in reference 1. This work on reprocessing is particularly important for the practice of reprocessing scrap material. As a consequence of the rotation of the screw during extrusion, processing generates high shear stress and strain conditions. PES samples became darker, changing from amber to black color after 6 processing cycles. All PES samples can be dissolved completely in dimethylformamide (DMF) and dichloromethane (DCM), indicating that cross-linking did not occur during reprocessing by extrusion.

The mechanical properties and the melt flow rate, MFR, of PES after different extrusion cycles are shown in Table 5. As can be seen, reprocessing causes a clear decrease in the MFR after processing. This indicates an increase in the molecular weight of the polymer. An increase in the molecular weight and viscosity due to processing has occurred. However, reprocessing does not cause a clear variation in mechanical properties.

Table 5. Effect of reprocessing on the properties of PES.

Extrusion cycles	Tensile properties					MAR. (g/10min)	
	Strength (MPA)		Strain (%)				
			Modulus				
	Yield	Break	(Gpa)	Break	Yield	2.16kg	5.0kg
0	---	---	---	---	---	1.82	3.28
1	81.2	56.6	1.93	55	10.0	1.54	2.78
2	77.3	56.0	1.67	42	10.2	1.68	2.82
3	76.3	55.2	1.42	21	11.2	1.27	2.71
4	73.3	51.7	1.20	23	11.3	1.26	2.78
5	73.1	52.3	1.24	37	11.7	11.7	0.94
6	75.0	52.5	1.27	35	11.6	0.82	2.48

Reprocessability of PES [17]

To ensure a high performance of the finished article, the material after reprocessing should have acceptable properties. As can be seen in Table 5, reprocessing at normal extrusion temperatures does not cause a clear variation in tensile properties, except a small decrease in modulus after 6 cycles. The PES samples after long kneading times have no observable change in structure and T_g, as can be seen in Table 4. All these results indicate that reprocessing changes only slightly affect the properties that are more related with practical use.

The MFR of PES after several extruding and cutting cycles decreases under different test conditions. As can be seen in Table 5, the MFR retention are 50% and 70% after 6 and 4 cycles, respectively. We also can see in Table 4 that the torque increases is less below the 40-min kneading time but it increases by 50% over the minimum torque after 50-min kneading times.

Based on the above fact, we can conclude that the viscosity of PES melt increases when it is subjected to prolonged heat and shear, and that the processing properties are affected. So, from the view of processing properties, the use of reground should be strictly controlled: it is recommended that no more than 4 cycles be used.

Thickening of Melt During Processing of PES

In technical circles, it is well known that the rheological properties of polymers change as a result of repeated application of shear, e.g., reground extruded polymer scrap does not behave in a manner identical to the virgin material. Under processing conditions, a polymer may experience thermal, thermo-oxidative, or mechanochemical degradation. A number of specific chemical reactions may also occur. Properties of plastics deteriorate mainly as a result of significant changes in polymer structure. In general, the reaction involves main chain scission, and this leads to a decrease in average molecular weight and viscosity of the melt polymer [30-32].

Earlier work has shown that the melt viscosity of PES rises as a result of repeated processing. The phenomenon known as thickening [9,17]. Similar phenomenon occurring during the processing of low-density polyethylene (LDPE) and high-density polyethylene (HDPE) are reported by Scott [33] and Pardon [30], respectively.

According to our work [17], it seems that the phenomenon of thickening is a result of reaction brought about by prolonged shear and heat/oxygen over a specified range of temperature. Reprocessing at normal extrusion temperatures does not cause a clear variation in mechanical properties. Calorimetry analyses also indicate that glass transition of all PES samples do not have distinct changes before and after processing. The glass transition point (T_g) remains constant irrespective of kneading conditions. The samples after processing were fully soluble in the solvent of DMF indicating the absence of cross-linking. Hence, there are not any notable effects on the structure and properties of PES samples after processing. The color of PES after kneading and extruding however does change. Processing causes a progressive darkening of the samples from an original amber color to brown, to a distinct black color. And the phenomenon of thickening is a good example of changes during the processing of PES.

Experimental Investigation

PES was passed through a screw extruder, solidified, and the rheological properties monitored afterwards using a capillary rheometer. The melt viscosity of PES extruded repeatedly is shown in Figure. 11.As shown, the viscosity of PES rises steadily with an increase in the number of extrusions. This is the phenomenon of thickening.

Obviously, the increase in melt viscosity comes about due to some sort of chemical modification. There is no reason to believe that the processes, whatever they may be, are caused exclusively by shear. Since, PES is inevitably kept hot for a long period of time because of the successive number of extrusions, it was decided to attempt to separate out any effect due to temperature alone the overall effect of heating and extrusion.

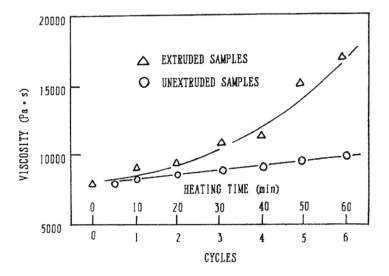

Figure 11. Melt viscosity of PES after repeated extrusion at   320°C
compared with that of samples kept for the same time.

To check the effect of temperature, PES was kept hot in the barrel of the MFR apparatus for a period of time corresponding to the residence time of PES during extrusion. The melt viscosity of PES rises slightly (Figure 11). It shows that the viscosity does indeed rise simply by leaving the PES hot for a prolonged period. It will be seen that the value after 60 minutes of heating is comparable to that after one extrusion but this would involve about 10 minutes of total heating. Thus, it is clear that shear and heating both contribute to the observed rise in melt viscosity. The phenomenon of thickening was caused by prolonged shear, heat and oxidation under processing temperature [17].

There are numerous experimental techniques that can be employed in the characterization of structure of PES [34-38]. A systematic study was conducted to investigate the changes of PES using nuclear magnetic resonance (13C NMR), Fourier transform infrared spectroscopy (FTIR) and X-ray photoelectron spectroscopy (XPS or ESCA) [9,39].

The IR spectra of PES from 4000 to 400 l/cm, were recorded on a Nicolet MX-1 FTIR spectrometer. All data indicate that it is not possible for PES to have a biphenyl structure. It was found that there is a small peak at 763 l/cm appearing in PES samples after processing. This peak becomes stronger with an increase in kneading time and with the number of extrusions. Clearly, branching occurs during the processing of PES.

The NMR spectra were obtained from a AC-80 type spectrometer. There does, however, appear to be a split in spectra of CI to double peak. A possible explanation of this change is that branching is taking place at the ortho-position on phenyl rings with respect to ether group (-0-).

XPS of sample PES were carried out with a XSAM 800 photoelectron spectrometer. It is shown from the results of the XPS, that a change in structure after processing occurs due to the C of C-C bond of the phenyl ring being affected. This represent a possible mechanism for melt thickening.

It can be seen from results above, the degradation of PES during processing is different from other polymers. As PES is being processed a melt thickening takes place and the molecular weight of PES increases (17). It deemed that branching takes place during processing and chain scission occurs simultaneously to a lesser extent.

During processing initial radicals formed due to scission of the polymer chain being subjected to shear. Consequently, these radicals undergo reactions and branching and cross-linking takes place, according to the following schemes:

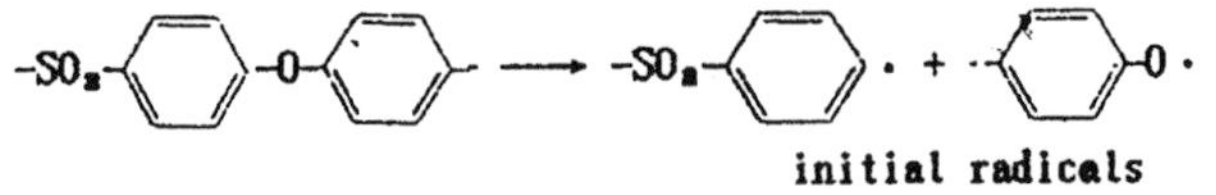

$\langle II \rangle$

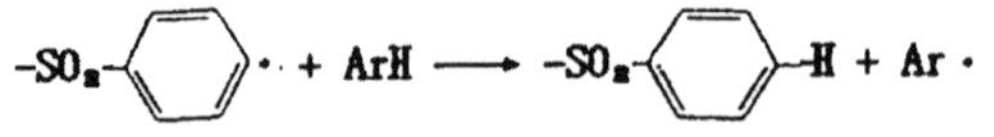

initial radicals

$\langle III \rangle$

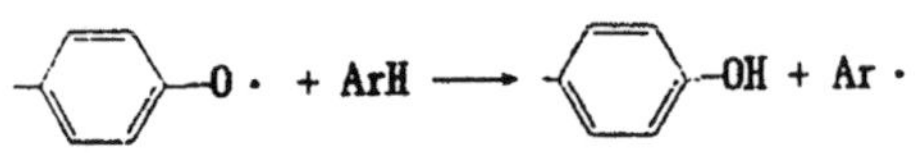

$\langle IV \rangle$

initial radicals + Ar · ⟶ branching $\langle V \rangle$

As shown, molecular weight increases as a result of branching $<V>$ and decreases by $<III>$ and $<IV>$. The general result would result in an increase in molecular weight and a widening of the molecular weight distributions of PES samples after processing.

Branching probably occurs by two schemes:

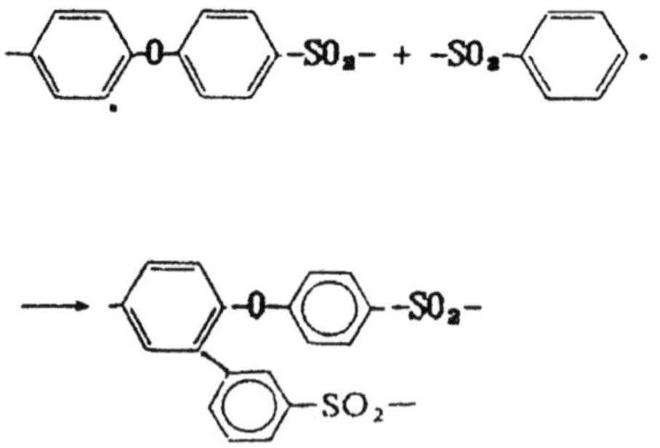

$\langle VI \rangle$

Reaction $<VII>$ which causes branching is supposed to proceed, but $<VI>$ would be negligible. Because from the discussion of the IR spectra, there is no biphenyl structure existing in PES samples after processing. And equation $<VII>$ can be a good explanation of the split in C1 peak of 13 C NMR measurement of PES being processed. In addition, the end groups of PES are easily scissioned under processing temperatures to form equation $<VII>$:

$$\text{C}_6\text{H}_4\text{—OCH}_3 \longrightarrow \text{C}_6\text{H}_4\text{—O} \cdot + \text{CH}_3 \cdot \qquad \langle\text{VIII}\rangle$$

Equation < VII > may a good explanation of the mechanism of branching result in thickening.

PES is sensitive to the processing conditions. As discussed, processing can result in chemical changes as a result of chain scission and branching reactions which can occur due to prolonged shear and heating under normal processing temperatures. Although the chemical modification occurring is not fully understood, it appears that branching takes place at the ortho-position with respect to ether group, and branching provokes an increase in molecular weight causing thickening of the melt. The interpretation presented, therefore, confirms that processing can result in modification of PES, so attention should be paid to this when studies on processing, synthesizing, and rheology of PES carried out. In view of the processability and of the current interest in re-using scrap, branching modification may well prove to be of importance. If these occurrences are considered disadvantageous, additives to "mop up" radicals would be necessary.

PRACTICAL ASPECTS OF PROCESSING

Injection Molding

Injection molding of thermoplastic resins is a well-known and widely practiced science. It constitutes a major processing technique for converting PES into a variety of end use products [41,42]. In basic terms, the process involves compressing and heating the molding pellets to convert to a molten resin, then transferring to a mold which is maintained under pressure until solidification occurs. Although PES resin has been molded in screw and plunger injection molding machines, in-line reciprocating screw machines are preferred. They offer more uniform plasticizing, a reduced injection pressure loss and less material hang-up areas.

After the specific processing characteristics of PES are considered, a successful molding operation can be accomplished.

Processing Profiles of PES and Drying Prior to Processing[1,3,4]

The first and most important requirement for obtaining high quality PES is to lower the moisture content of the pellets. If the PES molding resin is allowed to absorb excessive amounts of water prior to molding, the mechanical properties and appearance of the molded parts are impaired. Absorbed water can form steam that results in splay marks and internal bubbles in the molded parts.

Drying is a function of heat transfer and pellet permeability. The main variables affecting these requirements are pellet temperature and material residence time in the dryer. When using oven dryers, the resin should be spread in trays to a depth of approximately 25 mm, and then heated to 140-160°C for 3-4 hours. To eliminate any excessive heat history, it is recommended that the material be dried no longer than 48 hours.

It must be emphasized that the hopper and any open areas of the feed mechanism should be covered to protect the dried pellet from room atmosphere. If a hopper dryer is not available, then only a sufficient quantity of dry, heated PES pellets should be removed from the oven and placed in the hopper at one time. The time PES is exposed under these conditions before a harmful amount of moisture is absorbed could vary from 15 minutes to several hours, depending on the relative humidity.

Temperature Profiles [7,8,15]

Barrel temperatures are higher for PES than for most other thermoplastics. As a general guideline, barrel temperatures range from 280-320°C from rear to front. Generally, melt temperature should not exceed 350°C, and 320 C is preferred, The nozzle should act merely as a melt conveying pipe and should not affect the temperature of melt. The nozzle temperature should generally be about 320°C when molding PES. The rear zone temperature is particularly critical. Rear zone temperatures that are too low will not allow sufficient heat to be transferred to the polymer by conduction and will impose high torque loads on the screw drive system. This will result in erratic screw retraction problems. The high torque loads may even stall the screw.

Mold temperature is important too, particularly in determining final part finish and the molded-in stress level. Cold molds are more difficult to fill, necessitating high injection pressure and/or high melt temperature. Heated molds generally produce parts with a better finish and lower molded-in stress. Mold temperature of 120-140°C are recommended. At these temperatures, PES parts are ejected easily due to its high heat distortion.

Pressures and Cycle [15]

Generally, high molding pressures are necessary for molding PES. Normal injection pressure is 100 MPa, although good parts have been molded up to 200 MPa. Too low a pressure will result in parts that are not sufficiently well packed. Parts should be molded at the highest practical injection pressure. However, excessive pressure at the time of gate freeze-off can result in a gate area that is highly stressed.

The injection pressure and injection rate, although independently controlled, are interrelated. High injection rates are usually required to fill the part before it freezes, and in eliminating flow lines. This helps to create better surface finish.

Increasing back pressure increases the work done by the screw on the melt. This in turn, increases the temperature and its uniformity. Higher back pressure may be used to eliminate unmelted particles at a lower cylinder temperature.

The best cycle for PES resin calls for a quick fill, a hold time just long enough for the gates to freeze, and a brief cool period due to its high heat distortion temperature (HDT). PES can often be molded on faster overall cycles than other thermoplastics.

Table 6. Molding conditions for PES parts.

	Part A	Part B
Machine type	602	SZ-100
Temperature of cylinder °C		
Rear	280	280
Center	330	330
Front	335	335
Nozzle	325	320
Mold temperature °C	140	110
Injection pressure (gauge) (kg/cm²)	80	70
Drying	3 hours at 150 °C	

Cylinder Purging - Before and after handling MS resin, thorough purging of the barrel is essential. Since PES resin usually requires molding tempetatures above 300°C, well above the degradation temperature of most other plastics, it is imperative that all traces of other polymers be purged from the machine before running the PES resin. Complete purging is required because most other thermoplastics breakdown at the temperatures required for running PES, resulting in parts which neither look nor perform satisfactorily. The best purging materials are polypropylene(PP) and polyethylene (PE).

Mineral filled PP and PE can be used for purging after PES is introduced to the equipment. For best results, the initial purge should be followed with mineral filled PP or PE and dropping the cylinder temperature, and then switching to virgin PP or PE. With the cylinder heat still on, remove the nozzle and place it into an electric stove to burn out the PES. Thorough purging of the nozzle can be made by immersing it in organic solvent like DMF or DCM to dissolve traces of PES.

Extrusion - PES resin is not difficult to extrude. It does not demand extraordinary conditions or twin screw equipment, and standard techniques suffice.

PES resins, in form of pellets or powder, must be completely dried before processing. Generally, drying three to four hours at 140-160°C are recommended. Considerations like minimizing exposure of the processing resin to the atmosphere are necessary in order to limit moisture absorption. Minimum moisture pickup is especially important in extrusion as in injection molding. Since moisture absorption begins immediately with exposure to room air, pellets or powder should be processed within no longer than 15-20 minutes from removal from their containers.

Accurate control of cylinder temperature during extrusion of PES is necessary. Normal cylinder temperatures should be 280-320°C. For optimum processing performance and low torque loads on the screw drive system, it is important to set the rear zone at a higher temperature to allow sufficient heat to be transferred to the polymer by conduction, PES resin should be preheated. The rear zone temperature must not exceed a limitation that may result in premature melting of the polymer and bridging problems that result in erratic feeding. Extra care should be taken to minimize overall thermal exposure of both holdup time and upper processing temperatures.

The die and adapter temperature setting is particularly critical for successful extrusion of PES. Die temperatures that are too low will not allow molten PES to pass through the die easily and will result in freeze-off of PES in the die sometimes. If this happens the machine must be shut down, and the die should be removed and cleaned thoroughly. Die temperatures that are too high may result in darkening, surface frothing or degradating of the extrudate, on the other hand. Extrusion machines should have separate die and adapter controls. For the die and adapter, a proportional voltage feedback control is recommended.

Mechanical cleaning of an extrusion machine is relatively easy and with experience can be done in a short time. With the barrel heat still on, run PS or PP to clean out the PES. Then, remove the die and intermittently pull the screw, removing the hot melt and cleaning the screw with a brass brush.

The heated barrel can then be cleaned by a rotary brush with a extension rod. Sometimes there is no choice but to put the die and die adapter in DMF or DCM for thorough cleaning.

Strand Pelletizing

In injection molding and extrusion, use of strand pellet PES resin is certainly justified from a practical and economical point of view. The strand-cut pellet PES, strand extruded on single or twin-screw extruder using processing conditions listed in Table 7, were satisfactory in appearance and internal structure.

As mentioned above, a preferred die temperature of about 335°C is a key factor in strand extrusion. When die temperature exceeds 335°C, a cell structure forms in the strand. It is different from the cell structure resulting from moisture. The former has a closed cell structure whereas the latter is characterized by an open cell structure.

Pipe

PES pipe can be used in conveying hot air and ultra pure water in the electric and chemical industries [1,3]. Table 8 lists processing data recorded by Haake Sys40 at normal extrusion conditions.

Table 7. Processing conditions for strand extrusion of PES resins.

Extruder model	SLJ-56	SJ-20A/25
Screw	co-rotating twin	single
Diameter (mm)	56	20
Cylinder temp. °C		
rear	285	280
middle	290	295
front	295	300
Die temp. °C	300	315
Screw revolution (rpm)	14	10-50
Ampere of machine	4	1.0-2.2
Drying	3 hours at 160 °C	

Table 8. Typical processing conditions for pipe extrusion on an Haake Sys40 EGxtruder.

Torque (mg)*	Screw speed (rpm)	Die temperature (^{o}C)
4510	5	306
1939	10	316
5108	15	316

* Metric Conversion 1000mg = 9.807Nm.

ACKNOWLEDGMENT

The authors would like to thank Dr.Tieqi Li, Dr. Xiaoyi Gong and Dr. Qing Guan for useful discussions and Prof. Zhoujian Li for his support to this work.

REFERENCES

1. O.B.Searle, and R.H.Pfeiffer, Polym. Eng. Sci., 25 (3), 474 (1985).
2. E.Doring, Kunststoffe. 80(10), 1149(1900).
3. G.Blinne, M. Knoll, D. Muller, and K.Schlichting, Kunststoffe,75(1), 29(1985)
4. M. Kopiet, H. Zeiner, Kunststoffe, 77(10), 1020(1987).
5. M. Wolf, Kunststoffe, 79(2). 146(1989).
6. Z. Wu, Makromole chem., 171, 119(1989).
7. P. Friel, Kunststoffe, 77(4), 381(1987).
8. G. Blinne. Kunststoffe. 75(4), 219(1985).
9. Kun Qi, PhD. Dissertation, Chengdu University of Science and Technology, Chengdu, China, 1992.
10. Kun Qi, and R.Huang. Suliao (Chinese), 22(4), 34(1993). Chemical Abstracts 122:11306 (1995).
11. Kun Qi, and R.Huang, Inter. J. Polymeric. Mater., to be printed(1995).
12. D.R.Saini, and A.V.Shenoy, J. Elastomers Plastics, 17(3), 189(1985).
13. Donald V.Rosato, and Dominick V.Rosato, Plastics Processing Data Plastics Handbook, Van Nostrand Reinhold,New-York, 1990,Chap.10, pp 331-342.
14. R.Huang and Kun Qi, Suliao (Chinese), 22(2), 18(l993). Chemical Abstracts 121:232474(1994).
15. Kun Qi, S.Lai, Z. Zhang and R. Huang, China Plastics (Chinese), 6(2), 34 (1992).
16. Kun Qi and R.Huang, Suliao (Chinese), 23(4), 27(1994). Chemical Abstracts 123:57856(1995).

17. Kun Qi and R. Huang, Polym.-Plast. Techol. Eng., 33(2), 121-133(1994).

18. I. Mathew, K.E.George, and D.J.Francis, Intern. J. Polymeric. Mater, 21, 189(1993).

19. M.A.Grayson and C.J.Wolf, J. Polym. Sci., Polym. Phys., 25, 31(1987).

20. C. E, Browning, Polym. Ing. Sci., 18,16(1978).

21. C.Stacy, J.Appl. Polym. Sci., 32, 3959(1986).

22. P. Bonniau and A.R. Bunsell, J. Compos. Mater., 15, 272(1981).

23. H.G.Larter, and K.G. Kibler, J. Compos. Mater,, 10, 355(1976).

24. Z. Zhang, S.Lai, W. Wang and R. Huang,.China Plastics (Chinese), 4(2), 48(1990).

25. J.P.Critchley, G.J.Knight and W.W. Wright, Heat-Resistant Polymers, Plenum press, New York, 1983.

26. C. Arnold, J. Polym. Sci., Macromol. Rev., 14. 265(1979).

27. B. Crossland, G. J. Knight and W.W. Wright. Br. Polym. J., l8. 156(1988).

28. S. Kuroda, A. Hagura, and K. Horie, Eur. Polym. J., 25(6), 621(1989).

29. S. Kuroda. K. Terauch, K. Nogam and I. Mita, Eur. Polym. J., 25. 1(1989).

30. P. Pardon, P. J. Hendra and H. A. Willis, Plastics Rubber and Composites Processing and Applications, 20(5), 271(1993).

31. J. Muras, and Z. Zamorsky, Plaste und Kaut., 32, 302(1985).

32. K. B. Abbas, Polym. Ing. Sci., 20, 376(1980).

33. G. Scott, Polym. Eng. Sci., 24(13), 1007(1984).

34. A. Bunn, British Polym. J., 23, 307(1988).

35. C. Marletta, S. Pignataro, A. Toth, I. Bertoti, T. Szekely, and B. Keszler, Macromolecules, 24. 99(1991).

36. J. A. Gardella, S. A. Ferguson, and R.L.Chin, Applied Spectroocopy, 40(2), 224 (1986).

37. J. R. Brown, and J.H. O'Donnell. J. Appl. Polym. Sci., 23(9), 2763(1979).

38. J. R. Brown, J. Appl. Polym. Sci., 19, 405(1975).

39. Kun Qi, and R. Huang, Polym.-Plast.Technol.Eng., to be printed(1995).

4

Reactive Extrusion Processing of Elastomer Toughened Polyphenylene Sulfide.

Junzo Masamoto
Asahi Chemical Industry Co., Ltd.; Japan

INTRODUCTION

Polyphenylensulfide (PPS) features excellent mechanical properties, thermal stability, chemical resistance, flame resistance and precise moldability. However, PPS has a weak point of being a very brittle material. For use as electrical and electronic parts, automobile and mechanical parts, toughened PPS is desired. For these application, improving the toughness of PPS and developing an elastomer toughened PPS and related compounds are desired.

The targets of our development are as follows. The first point is to maintain the advantages of PPS such as its thermal properties, mechanical properties, etc. The second point is to improve the brittleness of PPS using the polymer alloy method.

REACTIVE PROCESSING OF ELASTOMER TOUGHENED PPS

Materials

Any commercially available PPS neat polymer is suitable for study. For example, the PPS neat polymer supplied by Tohprene Corporation (its trade name is "T-4") is suitable.

For the thermoplastic elastomer, one example is the commercially available olefinic elastomer composed of ethylene, acrylic ester and small amounts of maleic anhydride (-2 wt%). Another example is the ethylene propylene rubber (EPR) modified with a maleic anhydride group. One more example is the olefinic elastomer having a glycidyl group. These olefinic elastomers have the reactive functional group such as anhydride or an epoxide group.

Chemical Treatment of PPS

PPS was chemically treated with diphenylmethanediisocyanate (MDA) using the method described in our patent [1]. PPS was mixed with MDA, and extruded in a molten state using a twin screw extruder.

Reactive Processing

Various twin screw extruders can be used for the reactive processing. For example, BT 40 of Plastics Kogaku Kenkyusho, PCM 30 of Ikegai Tekko, and ZSK 40 of the Werner & Pfleidere Corporation were used in our development.

Elastomer Toughening

Figure 1 shows how PPS is toughened by elastomer alloying. The horizontal axis shows the flexural modulus of PPS and the vertical axis shows the notched izod impact strength.

The original PPS has a notched izod impact strength of about 1 to 2 kg-cm/cm, while elastomer-blended chemically, untreated PPS has a notched izod impact strength of about 6 kg-cm/cm. In these cases, elastomer blended PPS is thought not to be toughened.

Using MDA-treated PPS and an olefinic elastomer with a functional group of maleic anhydride, elastomer toughened PPS using the reactive processing method was developed.

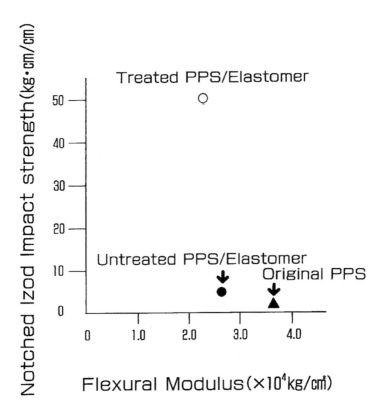

Figure 1. Effect of chemical treatment on toughness of PPS/elastomer alloy.

The elastomer alloy of the chemically treated PPS has a notch izod strength value of about 50 kg-cm/cm (2).

Though the notched izod impact strength of the original PPS is about 1 kg-cm/cm, the elastomer toughened PPS has a notched izod impact strength around 50 kg-cm/cm.

A notched impact fracture surface of PPS is observed using a scanning electron microscope.

Figure 2 shows the notched impact fracture surface of elastomer blended untreated PPS. The elastomer of the fracture surface is extracted with chloroform. The diameter of the extracted hole is about 1 μm. A scanning electron photograph of a notched impact fracture surface for a brittle PPS-elastomer blend shows a brittle surface.

Figure 3 is a photograph of the notched impact surface of the elastomer-toughened, chemically treated PPS. The elastomer of the fracture surface of PPS was extracted with chloroform. The diameter of the extracted hole is about 0.3 μm. A scanning electron photograph of the notched impact fracture surface for the tough PPS-elastomer alloy shows a tough fracture.

Figure 4 is a transmission electron photomicrograph of a microtomed section of toughened PPS. This case contains 20% by weight of the elastomer with similar particle size, and well-dispersed particles.

PROPERTIES

In the case of elastomer toughened PPS, there are two different application cases.

One case is unreinforced, and the other case is reinforced. Usually, we use our toughened PPS in the reinforced form. The mechanical properties of unreinforced and glass fiber reinforced elastomer toughened PPS are shown in Table 1. Although the notched izod impact strength of the general glassier reinforced PPS (Ryton R4) is about 7 kg-cm/cm, the glass fiber reinforced elastomer toughened PPS is about 22 kg-cm/cm.

Elastomer toughened glass fiber reinforced PPS maintains its original thermal properties. Heat deflection temperature (HDT) is shown in the table at the same level of over 260 °C.

Figure 5 shows the behavior of the dart impact strength of various glass

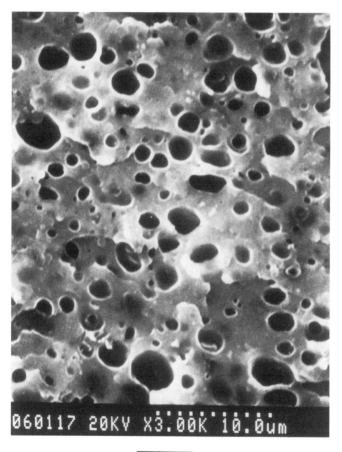

Figure 2. A notched fracture surface of elastomer/blended,
untreated PPS.

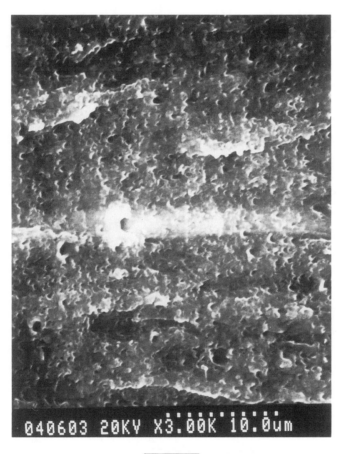

$$\overline{\qquad}$$
$$5\,\mu\,\text{m}$$

Figure 3. A notched impact surface of elastomer-toughened,
chemically treated PPS.

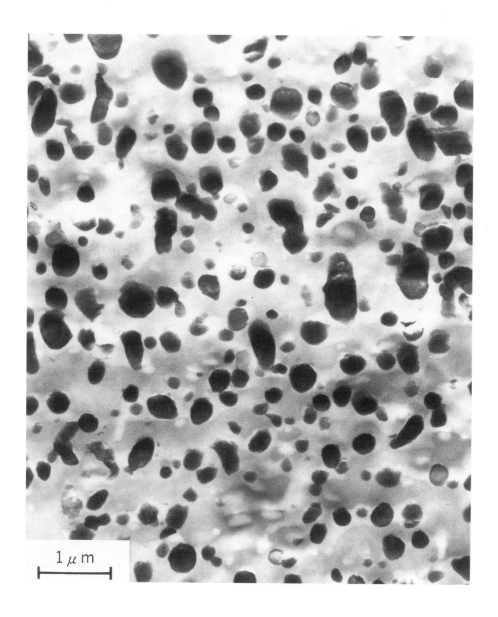

Figure 4. Microtomed section of toughened PPS.

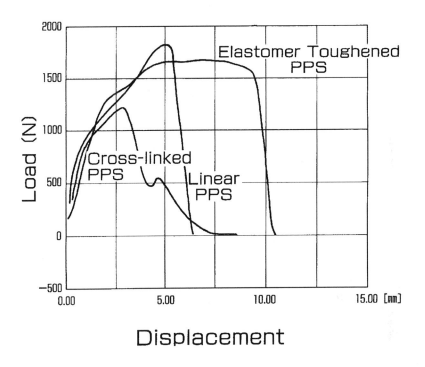

Figure 5. Dart impact properties of glass fiber reinforced PPS.

fiber reinforced PPS's. The horizontal axis is displacement, and the vertical axis is load. The area is energy for dissipation. Dissipation energies increased in the following order: cured and cross-linked PPS < linear PPS < elastomer toughened PPS. Elastomer toughened glass fiber reinforced PPS has an impact strength value twice that of the linear PPS and four times that of the cross-linked PPS.

Table 1. Properties of elastomer toughned PPS.

Item	Unit	Unreinforced PPS	GF 40% Reinforced PPS	Ryton R4* (Reference)
Tensile Strength	kg/cm^2	550	1,400	1,300
Elongation at break	%	25	3	1
Flexural Strength	kg/cm^2	800	2,200	2,000
Flexural Modulus	kg/cm^2	23,000	110,000	140,000
Izod Impact Strength (Notched)	kg•cm/cm	50	22	7
Heat Deflection Temperature (4.6 kg/cm^2)	^{0}C	150	>260	>260

* Properties are cited from catalog of "Dainippon Ink Corporation (DIC)-PPS".

Since this toughened PPS maintains the advantages of PPS, that is, thermal properties, mechanical properties, chemical properties and dimensional stability, toughened PPS has been in market development in the fields of electrical and electronics, automobile and mechanical applications, and good results have been obtained.

REFERENCES

1. Junzo Masamoto, Tetsuo Nakamoto and Kimihiro Kubo, US Pat. 5,191, 020 (1993) assigned to Asahi Chemical.
2. Junzo Masamoto and Kimihiro Kubo, Am. Chem. Soc. Polym. Prepr. 32(2), 215-216 (1991).

5

Processing of Nylon-3

Junzo Masamoto
Asahi Chemical Industry Co. Ltd.; Japan

INTRODUCTION

Breslow et al. reported that nylon-3 (sometimes described as nylon 3 or poly- β -alanine) is obtained using hydrogen transfer polymerization of acrylamide in the presence of an anionic catalyst:

$$CH_2 = CHCONH_2 \;\rightarrow\!\!\xrightarrow{anionic\ catalyst}\!\!\rightarrow -(CH_2CH_2CONH)_n-$$

Nylon-3, $(CH_2CH_2CONH)_n$, is a highly crystalline polymer with a high density of amide groups and is considered useful as a textile. The polymer melts at about 340^0C with considerable decomposition. Therefore, melt spinning cannot be used, but wet or dry spinning may be chosen for the fiber production.

Some properties of nylon-3 are thought to be different from those of other polyamides. Its glass transition temperature is supposed to be much higher than that of ordinary polyamides. Because the high density of the amide groups leads to a high absorption of water, nylon 3 may be expected to be similar to silk or cotton. Considering the chemical structure, Young's modulus of nylon-3 should be comparable to that of silk.

After the study by Breslow [1], various studies were completed [2-9]. At present, nylon-3 is commercially produced at Asahi Chemical and others, as an excellent stabilizer for polyoxymethylene.

PREPARATION OF NYLON-3

Various methods for the preparation of nylon-3 are reported in references and patents. Among the methods, the hydrogen transfer polymerization of acrylamide is very interesting from both the academic and industrial viewpoints.

Breslow et al.[1] found that when acrylamide was polymerized in the presence of a strong base catalyst, the polymerization occurred with hydrogen transfer, and nylon-3 was obtained.

$$CH_2 = CHCONH_2 \rightarrow \frac{t-BUONa}{} \rightarrow -(CH_2CH_2CONH)_n-$$

The polymerization was completed in the presence of a strong base catalyst (e.g. t-BuONa, t-BuOK) using inactive solvents such as toluene, pyridine, chlorobenzene and o-dichlorobenzene from 80 to 200 ^{0}C.

The initiation and propagation reactions seem to be effected by polymerization conditions such as the type of catalyst and type of solvent, but there are many unclear points [10,11].

Masamoto et al. [5] polymerized acrylamide in dioxane using n-BuLi as the catalyst. As shown in Figure 1, the polymer yield was nearly 90 % soon after the start of the polymerization. However, the reduced viscosity of the polymer gradually increased with the polymerization time. Furthermore, as shown in Figure 2 the primary amide and double bond of the reactants decreased with the polymerization time. These results indicate that the deforming factor for the molecular weight of the polymer is the polymer reaction between the double bond and primary amide. Similar results were also reported by Camino et al [6]. Hung [12] also discusses the importance of the vinyl group formation within the obtained polymer.

In the base catalyzed polymerization of acrylamide, the vinyl polymerization is reported by Nakayama et al. [7] to occur in addition to the hydrogen transfer polymerization process.

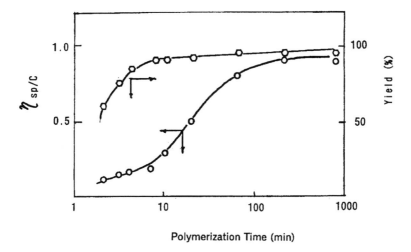

Figure 1. Yield and reduced viscosity (η_{sp}/C) of the polymer
vs.polymerization time [5].

Polymerization Temp: 120 $^{\circ}$C, Solv.Dioxane,
Cat.: n-BuLi, $(M)_o$: 1.4 mol/l, $(M)_o/CaT$: 30/1 (mol/mol)

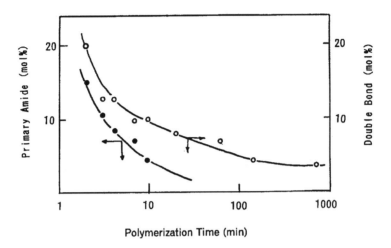

Figure 2. Double bond and primary amide
vs. polymerization time [5].

Sol. Dioxane,   Polymerization temp. : 120°C,
Cat.: n-BuLi, (M)o: 1,41mol/l, (M)o/Cat.: 1/30

If we hypothesize that the hydrogen transfer polymerization product of acrylamide is the polymeric structural material of nylon-3, then the following two points arise as potentially serious problems.

1) Vinyl polymerization readily occurs.

2) A high molecular weight polymer is difficult to obtain.

Considering these two points, Masamoto et al. [5,8] discussed how to obtain a high molecular weight nylon 3 which does not contain vinyl polymerization. They found that when the polymerization catalyst (t-BuOK) is finely dispersed in the polymerization solvents, the vinyl polymerization does not easily occur. Moreover, they found that when acrylamide is anionically polymerized in the presence of an inorganic salt, a high molecular weight nylon-3 is obtained. To finely disperse the catalyst, an alcoholic solution of alcoholate (mainly, t-BuOK) was mixed with high boiling point polymerization solvents (mainly, o-dichlorobenzene), and the alcohol was then evaporated under reduced pressure. An inorganic salt such as potassium chloride was added to the finely dispersed alcoholate-o-dichlorobenzene solution, and the acrylamide was polymerized. A high molecular weight nylon-3 with less than 2% vinyl polymerization unit was obtained. In Figure 3, the polymerization results using t-BuOK as the catalyst which was finely dispersed in o-dichlorobenzene and KCl as the inorganic salt are shown. The reduced viscosity of the polymer was increased with the addition of the KCl. The T- ratio, which was defined as the ratio of hydrogen transfer polymerization to vinyl polymerization, did not change but remained constant. The effect of the inorganic salt is thought to increase the reactivity of the growing polymer chain [8].

Morgenstern and Berger also reported that a nearly pure nylon- 3 was produced using the hydrogen transfer polymerization of acrylamide [13]. They confirmed these results using an NMR method.

PROCESSING OF NYLON-3 FOR FIBER FORMATION [14-17]

Nylon-3 melts at about 340°C with considerable decomposition. Therefore, melt spinning cannot be used, but wet or dry spinning may be chosen for the fiber preparation.

Solubility of Nylon -3 [14]

Nylon-3 is insoluble in most organic solvents and in water, as well as in aqueous basic solutions and dilute acid solutions. However, nylon-3 is soluble in water above 140-170°C, strong inorganic acids, organic acids, some aqueous and methanolic inorganic salts, hot aqueous phenol, and aqueous chloral hydrate. As for inorganic salts, $Ca(SCN)_2$, $LiCl$, $CaBr_2$,

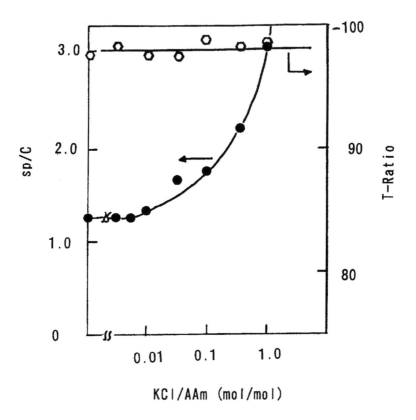

Figure 3. Effects of the KCl on the polymerization [8].

Solv.: o-Dichlorobenzene, Cat.:t-BuOK,
(M)o: 1.4mol/l, (M)o/Cat.: 30/1(mol/mol), 115°C

ZnCl$_2$, and SbCl$_3$ are effective solvents. Among these solvents, only formic acid is available as a useful spinning dope for the production of the nylon-3 fiber. The polymer concentration in the spinning dope using formic acid as the solvent is from 30 to 40 wt% and the viscosity of the spinning dope is comparably low. Its viscosity is about 1/50 of the dichloroacetic acid solution. This means that the nylon-3 macromolecular chain shrinks in the formic acid solution. The concentrated formic acid solution of nylon-3 with a polymer concentration above 38 wt% tends to gel with time. The gelation phenomenon is strongly related to the polymer concentration. If the polymer concentration is over 45 wt%, then the nylon-3 formic acid solution instantly exhibits gelation. This gelation is thought to be caused by the poor interaction between the nylon-3 and formic acid. Gelation is not desirable for preserving the spinning dope, but the gelation of the formic acid solution is a necessary factor for the fiber forming process.

Wet Spinning of Nylon- 3 [16,17]

Wet spinning of nylon-3 is only possible when the spinning dope uses formic acid solution. Masamoto et al. [17] studied various solvents, and they established these conclusions. The available reduced viscosity of the polymer was from 0. 9 to 3. 0, and the preferable polymer concentration is from 38 to 40 wt%. In wet spinning, the combination of the solvent and coagulant is important. The coagulants which exhibit fiber forming abilities are as follows:

Organic solvents such as propanol, i-propanol, butanol, hexanol, diisopropylether, dioxane, tetrahydrofuran, methyl acetate, ethyl acetate, i-propyl acetate, butyl acetate. Water and water containing formic acid were not useful coagulants.

The coagulants which demonstrate excellent fiber forming abilities are thought to act as a base for the formic acid, and act as a base or be neutral to the nylon-3. This indicates that the coagulants have an affinity for the formic acid which is the solvent for the nylon-3, and they do not have an affinity for the nylon-3. A typical nylon-3 fiber forming process is shown in Figure 4. From these results, formic acid is diffused into the coagulants during the fiber forming process. However, little penetration of the coagulants into the protofiber was observed. Thus, the fiber formation of nylon-3 is characterized by desolvation and deswelling, contrary to that of other polymers in the wet spinning process. Because of the desolvation of nylon- 3 protofiber, the polymer concentration is increased in the nylon-3 spinning dope, therefore, gelation occurrs inside the protofiber and a gel led protofiber is formed. According to desolvation, a protofiber with a dense structure is formed. The X-ray diffraction patterns of the protofibers shows that the fibers are already crystallized. Thus, the molecular orientation of nylon-3 is thought to be quite difficult, considering the high density of the amide groups.

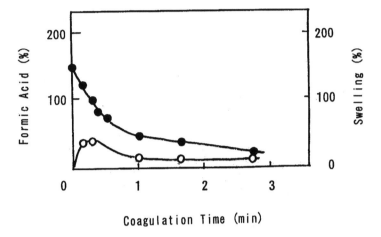

Figure 4. Fiber-forming process of nylon-3 in wet spinning, using formic acid as the solvent and ethyl acetate as the coagulant [17].

Swelling (%): "(Weight of protofibers)/Weight of polymer in protofibers)"x100

Dry Spinning of Nylon-3 [15]

Dry spinning of nylon-3 using formic acid needs a higher molecular weight polymer than wet spinning. With the polymer having a weight average molecular weight of 90,000 (reduced viscosity: 0.9) the filaments cannot be continuously wound. However, a polymer with a weight average molecular weight of 240,000 (reduced viscosity: 1.8) can be spun dry and filaments can be continuously wound. In the dry spinning of nylon- 3, the spinnability of the spinning dope is an important factor for the fiber forming ability.

Drawing of Nylon -3 Fibers [17]

Considering the strong inter- and intramolecular interaction of nylon-3, the stretching of the fibers in a state of lowered transition points of the polymer was thought plausible. From this viewpoints, the stretching of the fibers that contain a large quantity of formic acid or stretching of the fibers in such media as hot water and glycerine, which are thought to have some interaction with the polyamides, was thought plausible. By varying the coagulation time, nylon-3 fibers containing 20-30 wt% formic acid were obtained, but stretching these fibers at room temperature was impossible. The stretching of the undrawn fibers in such media as hot water, hot steam, and hot glycerine was also investigated. However, the nylon-3 fibers could not be stretched in these media. In such media as silicone oil, Wood's metal, and hot air, which are thought to have no interaction with polyamides, the nylon-3 fibers could be stretched above their glass transition temperature $(170\text{-}180^0C)$.

The effects of the drawing temperature and formic acid content in the fibers on the maximum draw ratio were investigated. The most favorable drawing temperature with the maximum draw ratio of the undrawn fibers decreased with an increase in the formic acid contents in the undrawn fibers. The X-ray diffraction pattern of drawn the nylon-3 fibers is shown in Figure 5. It revealed good molecular orientation. Birefringence of the most oriented fibers was increased to a value of $8\text{x}10^{-2}$, which is a higher value considering the fact that ordinary polyamides showed a value of $6\text{x}10^{-2}$. This was due to the high amide concentration.

PROPERTIES AND STRUCTURE OF NYLON-3

Various works on the properties and structure of aliphatic polyamide have been published. However, very little work has been completed on the aliphatic polyamides of nylon-3.

Figure 5. X-ray diffraction pattern of drawn nylon-3 fibers.

Crystal structure of Nylon- 3 [18-21]

The crystal structure of nylon-3 was determined by Masamoto et al.[18] using the X-ray diffraction pattern. The foure crystal structures of nylon-3 have the designated modifications: I, II, III, and IV.

The crystal structure of the drawn nylon-3 fibers (modification 1) was determined using X-ray diffraction. This structure is similar to the alpha-crystal of nylon-6 and nylon-4, and the unit cell is monoclinic. The unit cell is as follows:

a=9.33 A, b=4.78 A (fiber identity period), c=8.73 A, β=60^0, and the space group is P_{21}.

In this structure, the b-axis was found to be normal to the basal plane. The theoretical density for nylon-3 with four monomeric units in a unit cell is 1.39 g/cm^3, and the density was determined in a density gradient column of toluene and carbon tetrachloride to be 1.33 g/cm^3 at 25^0 C.

The identity period along the fiber axis shows an extended planar zigzag conformation with a plane of amide groups normal to the fiber axis. This is consistent with an antiparallel arrangement, that is similar to that found in nylon- 6. A displacement of the hydrogen bonded sheets along the b-axis by three fourths of the b-axis results in more efficient packing of the CH_2 group in the unit cell. The NH-O hydrogen bond distance was calculated to be 2.88 Angstroms, and the distance between the sheet was 3.78 Angstroms.

It is interesting to note that nylon-3 has a monoclinic unit cell which is unlike other ordinary odd-numbered nylons such as nylon-7, -9 or -11, which display a triclinic or pseudohexagonal unit cell. The amide concentration in the polymer might be a deteremining factor in the formation of a monoclinic, triclinic, or pseudohexagonal structure.

Three additional crystalline forms of nylon-3 were observed [19] and were designated modifications II, III and IV. These four modifications can be transformed mutually [18]. Similar results were also confirmed by Munpoz-Guerra et al. [19].

It was generally known that long chain polyamides like nylon-6 have the γ-type crystal in which the amide group of the main chain is torsioned [22]. However, according to the results of X-ray diffraction and infrared spectrometry, the formation of the γ-type crystals in nylon-3 is not observed even after treatment with iodine and hydrogen chloride gas. Therefore, polyamide with its short sequences of methylene groups are believed to contain no γ-type crystals.

Properties of Nylon- 3 [11,17]

In Figure 6, the stress-strain curves of the drawn nylon-3 fibers are shown. Although drawn nylon-3 fibers are brittle, a wet heat treatment improved the tensile elongation properties, and the mechanical properties were as follows: tensile strength, 2-3 g/d; tensile elongation, 10-20 %; Young's modulus, 800-1200 kg/mm^2. Young's modulus was quite high comparable to that of silk.

Because nylon-3 had a high amide concentration, a high moisture regain was expected. In Figure 7, the moisture regain curve of nylon-3 fibers is shown compared to silk. At a relative humidity of 65 %, the undrawn nylon-3 fibers have a moisture regain value of 9 %, which is quite similar to that of silk. Furthermore, the moisture regain curve was also quite similar to that of silk.

In Figure 8, the viscoelastic properties of the nylon-3 fibers are shown [17]. The temperature showing the maximum tan d, which might be attributed the temperature of the hydrogen bonds scission in the amorphous region, was approximately 70^{0}C higher than that of ordinary polyamide. In Table 1, the miscellaneous properties of nylon-3 are summarized and compared to those of silk and nylon-6. The temperature that exhibits the maximum tan delta is 180^{0}C, and is very high compared to nylon-6. The moisture regain value is 7 %, and this value is comparable to that of silk. Furthermore, Young's modulus was 700-1200 kg/mm^2, which was much higher than that of other synthetic fibers. These characteristic properties of the nylon-3 fibers indicated that the nylon-3 fibers can be said to be essentially similar to silk.

Table 1. Miscellaneous properties of nylon-3 fibers [11].

Properties	Nylon-3	Silk	Nylon-6
Melting point (^{0}C)	340	---	215
Density (g/cm^3)	1.33	1.37	1.14
Tensile Strength (g/d)	2-3	3.3-5.5	4.1-5.8
Tensile Elongation (%)	10-20	13-25	26-40
Young's Modulus (kg/mm^2)	800-1200	650-1200	80-250
Moisture Regain (%)	7	9	4-4.5
Temp. Max. Tan d (^{0}C)	180	---	110

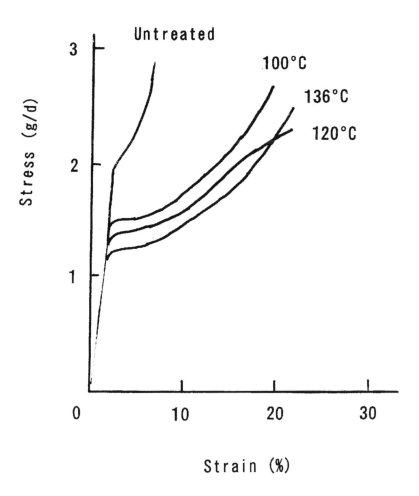

Figure 6. Effects of wet heat treatment of stress-strain curve of nylon-3 fibers [11].

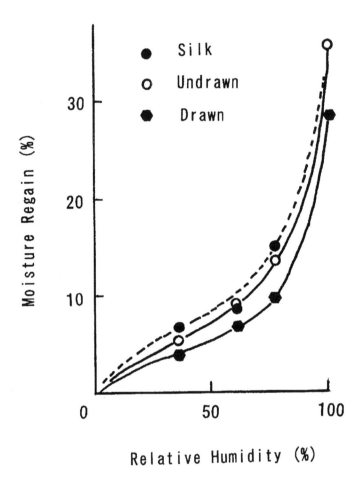

Figure 7. Moisture regain curve of nylon-3 fibers [11].

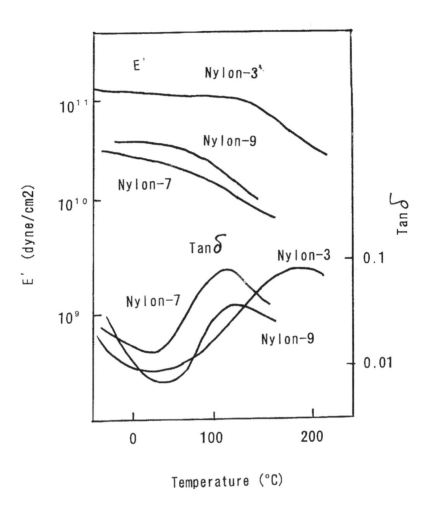

Figure 8. Viscoelastic properties of nylon-3 fibers [17].

COMMERCIAL APPLICATIONS OF NYLON-3

Although the nylon-3 fibers exhibited properties quite similar to that of silk, the nylon-3 fibers have not yet been commercialized. At present, various applications of nylon-3 are being investigated, but the only commercially available application is as a stabilizer for polyoxymethylene.

As the stabilizer for polyoxymethylene, the copolyamide (nylon 6-66-10) was known to be a formaldehyde scavenger. Recently, the Asahi Chemical researchers determined that nylon-3 was an excellent stabilizer for polyoxymethylene2 [3]. Because nylon-3 has a high amide concentration, it appeared to be an excellent formaldehyde scavenger. At present, nylon-3 is commercially produced only at Asahi Chemical as the stabilizer for polyoxymethylene. Several proposals for the production of nylon-3, which is suitable for the stabilization of polyoxymethylene, have also been proposed [24].

An advantage of using nylon-3 as a stabilizer for polyoxymethylene is that nylon-3 exhibits excellent thermal stability because of its high amide concentration. Furthermore, a polymer composition which contains nylon-3 shows negligible decoloration when the polymer melt remains in an injection machine for a long time. Also nylon-3 does not deposit onto the surface of the injection mold because of its high melting point. These three points are the characteristic merits of nylon-3 for this application.

REFERENCES

1. D. S. Breslow, G. E. Hulse, A. S. Matlack: J. Am. Chem. Soc., 79, 3760 (1957); A. S. Matlack: US Pat. 2,672,480 (1954), assigned to Hercules Powder Co..
2. N. Ogata: Bull. Chem. Soc. Japan 33, 906 (1960); J. Polym. Sci., 46, 271 (1960); Makromol. Chem., 40, 55 (1960).
3. H. Tani, N. Oguni, T. Araki:Makromol.Chem., 76, 82 (1964).
4. L. Trossarelli, M. Guaita, G. Camino: ibid., **105**, 285 (1967).
5. J. Masamoto, K. Yamaguchi, H. Kobayashi: Kobunski Kagaku, 26, 631 (1969).
6. G. Gamino, M. Guaita, L. Trassarelli: Makromol Chem., **136**, 155 (1970);J.Polym.Sci.,Lett.Ed., **15**, 417(1977); G. Gamino, S. L. Lim. L. Trossarelli:Europ.Polym.J. 13, 473 (1977); G. Gamino, M. Costa, L. Trassarelli: J.Polym. Sci., Chem. Ed., 18, 377 (1980).
7. H. Nakayama, T. Higashimura, S. Okamura: Kobunski Kagaku, 23, 433 (1966); 23, 439 (1966); 23, 537 (1966); 24, 427 (1967).
8. J. Masamoto, C. Ohizumi, H. Kobayashi: ibid., 26, 638 (1969).
9. J. Glickson, Y. Applequist: Macromolecules, 2, 628 (1969).
10. T. Ohtu: Kagakuzoukan, 53, "New Polymer Synthesis" (Ed. by T. Ohtu and K. Takemoto), P. 169, Kagakudojin(1972).

11. J. Masamoto: Poly- β-alanine Fiber, " Polyaminoacids - Application and Survey" Ed. by Y. Fujimoto, p. 234, Kodansha(1974);Kobunshi, 24, 187 (1975).

12. S. Y. Huang: ACS Polym. Prepr. 24(2) 60(1983); S. Y. Houng, M. M. Fisher: ibid. 34 (l), 997 (1993)

13. U. Morgenstein, W. Berger:Makromol.Chem. **193,** 2561 (1992).

14. J. Masamoto, Y. Kaneko, K. Sasaguri, H. kobayashi: Sen-i Gakkaishi 25, 525 (1969); J. Masamoto, K. sasguri, C. Ohizumi, H. Kobayashi: ibid., 26, 239 (1970).

15. J. Masamoto, K. Sasguri, H. Kobayashi: ibid., 26, 246(1970).

16. J. Masamoto, K. Yamaguchi, H. Kobayashi: ibid., 25, 533 (1969); J. Masamoto, C. Ohizumi, H. Kobayashi: ibid., 26, 16 (1970); J. Masamoto, A. Miyake, C. Ohizumi, H. Kobayashi: ibid., 26, 102 (1970). Masamoto, C. Ohizumi, H. Kobayashi: ibid., 26, 138(1970).

17. J. Masamoto, K. Sasguri, C. Ohizumi, K. Yamaguchi, H. Kobayashi: J. Appl. Polym. Sci., 14, 667 (1970).

18. J. Masamoto, K. Sasguri, C. Ohizumi, H. Kobayashi: J. Polym. Sci., A-2 , 8, 1703 (1970).

19. S. Munoz-Guerra, J. M. Fernandez, A. Rodriguez-Galan, J. Subirana: J. Polym.Sci., 23 733 (1985)

20. J. Masamoto, H. Kobayashi: Kobunshi Kagaku, 27, 220, (19700.

21. J. Masamoto, Y. Kaneko, H. Kobayashi: ibid., 27, 301, (1970).

22. Y. Kinoshita: Makromol. Chem, 33, 1, (1959).

23. G. Yamamoto, T. Misumi: US Pat. 4, 855, 365 (1989); US Pat. 5,015,70 7 (1991.), assigned to Asahi Chemical.

24. H. Hazawa, J. Ohtake: Japanese Exam. Pat. 5-47568 (1993); J. Masamoto: Japanese Pat, under Application, assigned to Asahi Chemical.

6

Pultrusion Processing

S.M.Moschiar, M.M.Reboredo, A. Vazquez
Institute of Materials Science and Technology (INTEMA)
Mar del Plata, Argentina

INTRODUCTION

Pultrusion is a continuous process whereby reinforcing fibers are first pulled through a resin bath, then into a shaping and forming guide system and finally into a die where the product is generally heated and cured to its final dimensions.

Uses

Pultruded profiles have participated competitively in markets like: electrical, corrosion, construction, transportations, as well as aerospace and defense. Figure 1 shows some of the typical profiles manufactured by this process.

The products fabricated by this method are (1): transformer air duct spacer sticks, ladders, bus bar supports, fuse tubes, cable support trays, fishing rods, antennas, skate boards, tool handles, ski poles, golf shafts, bridges and platforms, stairs, pipes and tubes, leaf springs, seating, bus luggage racks, etc.

Raw Materials

The majority of the profiled products are produced with fiber glass reinforcement and thermoset polyester or vinyl ester resins.

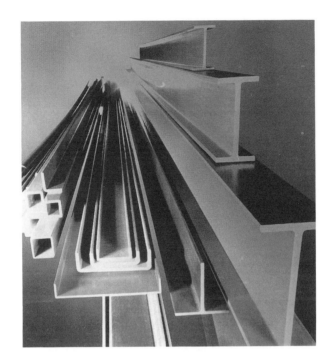

Figure 1. Pultruded profiles.

Resins as epoxies with low viscosity (2) are used and they generally present less shrinkage value than polyester resins.

Methyl methacrylate resins are being used. These reinforced resins are pulled at speeds 2 to 3 times faster than conventional polyesters.

Three pultrusion grades of phenolic resin systems are used (3-4) and they provide high flame retardancy, plus low smoke ant toxic gas emission in fires.

Thermoplastic resins as polypropylene (PP) (5) and poly phenylene-sulfide (PSP) (6) are being used as matrices for pultrusion process, also.

The material reinforcements are glass fiber in the form of continuous roving, continuous strand mat and cloth. Carbon and aramid fibers have also been incorporated in profiles (7).

Other thermoplastic fibers as nylon fibers, high density polyethylene fibers and polyester fibers are also being investigated for use in pultrusion.

Formulations can include fillers such as calcium carbonate, kaolin clay and alumina trihydrate. Internal mold release agents such as fatty acid or stearate soaps and pigments are also some of the materials used in typical formulations.

Process Description

Pultrusion is a method for producing continuous lengths of fiber-reinforced parts. A typical pultrusion machine consist of six in-line stations: a) filamentar raw material dispensing creels, b) resin impregnation tank, c)excess resin removal devices, d) heated die zone, e) gripping/pulling device, and f) cutoff saw.

Reinforcing fibers are first pulled from creels where unidirectional and multidirectional reinforcements are organized prior to going into a resin bath. Fiber roving dispensed from center-pull packages sitting on bookshelf-type racks. Fiber mats, supplied in large rolls, are pulled from the outside tangent from simple spindles. Occasionally, roving is dispensed tangentially from spindles where elimination of roving twist is a consideration in the end product. Most other reinforcements and surfacing materials, as carbon fiber and nonwoven veils are dispensed tangentially.

The reinforcement is impregnated with a resin matrix and precisely formed just prior to entering a heated die (Figure 2). The steel die is machined to the final dimensions of the part desired. A continuous pulling system alternately clamps and pulls on the part, pulling the material from the creels through the resin station and the forming guide area into the die and ultimately through a cut-off saw which automatically cuts the part to the

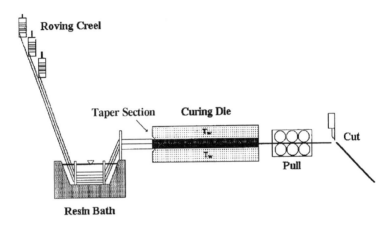

Figure 2. Scheme of the pultrusion system.

desired manufactured length.

The delivery of heat into the material prior to its entering the die is accomplished via Radio Frequency (RF) preheating, induction heating or by conventional conductive heating of the resin. Desirable characteristics such as increased process rates and the reduction of thermal stresses induced by gradient of temperature and less content of voids can be achieved by the heating methods listed over more conventional approaches (8).

Pultrusion can only produce straight, constant shape products, or profiles. However, there are now some developments in the pulforming process for manufacturing curved or varying area/changing shapes (9)

UNIFIED APPROACH TO THE PULTRUSION PROCESS

The process design consist of the solution of the master model and their application in the analysis of different case studies of the pultrusion process under different processing conditions in order to obtain general criteria for the optimization of the main process variables, scheduling of the temperature or pulling velocity and material management as formulations of content and types of resin and fibers.

The master model consist of five submodels: a) Kinetic submodel or crystallization submodel; b) Heat transfer submodel; c) Pressure model; d) Pulling force submodel; e) Residual stress submodel. Figure 3 shows the generalized scheme of the pultrusion process.

In order to construct the master model, it is necessary to have the following approaches:

a) Chemical kinetics: which permits the description of the polymerization and cross linking reactions for thermoset resins. The evolution of the degree of reaction as a function of time at different temperatures is used to construct the kinetic submodel.

b) Crystallization kinetic: for the case of thermoplastic resins, the crystallization kinetic permits the evolution of the crystallization degree in the function of the time for different temperatures.

c) Chemorheology: The rheological behavior of reacting systems is governed by the viscosity which reflects the molecular structure induced by the cure reaction and the variation of the temperature mobility determined by temperature changes.

d) Heat Transfer: The thermal balance and the different conditions at the wall must be used in order to solve the heat transfer submodel in the die.

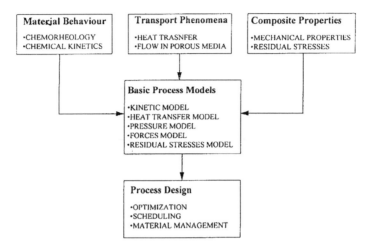

Figure 3. General approach in Pultrusion Process.

e) Flow in porous media: Darcy's law is usually applied to describe the conditions in the taper section. The aligned fibers are taken as cylindrical tubes between which the resin flow.

f) Mechanical Properties: The resin and fiber used as well as the content of fiber depend on the mechanical properties desired.

g) Residual Stresses: The temperature profiles on the heat die section and the pulling force determined the residual stresses in the final composites.

PULTRUSION OF THERMOSET RESINS

Generally the most suitable resin for a pultrusion process must possess the following properties:

1.- a suitable viscosity (0.5-2 Pa.s) in the impregnation tank to insure good fiber wet out and to avoid resin starvation when entering the mold.

2.- a long pot life in the impregnation tank to provide sufficient time for fiber wet-out and subsequent pultrusion processing.

3.- a high reactivity (thermosets), otherwise, short processing times will be not achieved.

4.- good wetting ability between fiber and resin. This means that the fiber bundles must be distributed evenly in the polymer matrix in order to obtain composites with the best mechanical and thermal properties.

Long pot life in the impregnation tank means that the resin viscosity only increases slightly for a long processing time; hence, the fibers have adequate time to be wetted out and the pultrusion process can be operated for a long time. However, the resin should have high reactivity so that the composite can be fabricated in a short time in the pultrusion die. The good fiber wet-out should be obtained to reach the optimum mechanical properties for the composites.

The following criteria, referred to as the matrix behavior, can be applied to the optimization of the pultrusion process:

a.- a final degree of reaction greater than 0.9 is required. Sometimes a postcuring step may be necessary to ensure complete reaction and to drive off volatiles from the composites. Postcure will increase the degree of crosslinking in the polymer matrix and, hence, may improve the mechanical properties of composite materials.

b.- the exothermic temperature peak must be minimized to avoid matrix degradation and strong gradients of degree of reaction and viscosity.

c.- the gel point must be reached close to the end of the die.

d.- the pulling rate must be maximized in order to achieve the highest product rate.

The maximum temperature and conversion profiles curves along the die length are moved to the entrance as the pulling velocity and die thickness decreases and die wall temperature increases. The degree of cure increases with decreasing pulling velocity and die thickness. These aspects are very important for the final performance of the part.

Experimental Measurement of Pulling Force and Pressure

There are few published data for an epoxy matrix in a pultrusion machine. Hunter (10) have shown results with an epoxy resin (EPON Resin 9310/ EPON Curing Agent 9360/Accelerator 537) with glass fibers. Additionally, Sumerak (II) performed experimental work with a polyester-glass fiber system in which he obtained temperature profiles, pressure and pulling forcesfor catalyzed and uncatalyzed polyester formulations at several filler loading. The pulling force measured with catalyzed resin was higher than without catalyst due to adhesion and friction of the cured resin at the wall.

Kinetic and Chemorheologic Submodels

A complete study of the different kinetic and chemorheologic models was done by Lee (12). This study reviewed the knowledge of material characterization in liquid composite molding.

Also, Kenny and Opalicki (13) reviewed the empirical kinetic model for thermoset resins and applied the master model for autoclave processing of epoxy resin. Different rheological models were also explained.

Tajima and Crozier (14) used the William-Landel-Ferry equation for the variation of epoxy viscosity during cure reaction. The epoxy resin studied was a special resin for pultrusion processing.

Heat Transfer Submodels

Han et al. (15,16) developed a mathematical model for pultrusion of unmatured polyester and epoxy resins considering a variable wall temperature along the die. They used only one resin - fiber ratio and the influence of pulling force on temperature and conversion profiles was discussed.

NG and Mana-ZIoczower (17) applied a mechanistic kinetic model based on the concept of free radical polymerization and corrected for diffusion controlled reaction of polyester resin in order to study the pultrusion temperature profiles in the pultrusion process. The presence of filler and fiber did not affect the kinetics of reaction and only acted as a heat sink for the composite system.

Trivisano et al. (18) developed a mathematical model for the pultrusion of epoxy-carbon fiber composites which was applied to the description and optimization of process variables to determine the best set of processing conditions (die temperature and pulling velocity as function of the composite thickness). They concluded that different heating zones must be used in order to improve processing, especially for thick composites.

Batch (19) developed and used a heat transfer model with pre-heating and convection cooling in a polyester- glass fiber system in a commercial pultrusion machine.

Vallo and Vazquez (20) used a heat transfer model to analyze the influence of process variables and material parameters (pulling velocity, wall temperature, fiber fraction) on the processing of an epoxy system (DGEBA+DGBD) cured by 3DCM amine with glass fiber.

Heat Transfer, Pressure and Pulling Force Submodels

Price (21) analyzed the thermorheological aspects of the processing of epoxy-carbon fiber and polyester-glass fiber. He used a heat transfer model for pultrusion and examined two limiting cases: i) isothermal case with uniform die wall temperature and ii) adiabatic case where heat conduction was considered negligible. However no correlation between process variables and temperature profiles was discussed.

Batch (22) presented a pulling force model in order to match theoretical results with Sumerak's experimental data. This work was useful for understanding pultrusion processing, since he was one of the first to model the pressure profiles and pulling forces in themoset resins. He also studied fiber compactation in the pultrusion taper section as well as taper shape.

Moschiar et al. (23) modeled pressure evolution in the die and the required pulling force. The heat transfer and pressure submodel were similar to Batch's model (22) however the pulling force submodel was different. It was considered only frictional solid-solid coefficients and it stated a condition for the pulling force calculation. The investigators compared their model with experimental results reported by other authors on the pultrusion of unsaturated polyester matrix-glass fiber composites.

Moschiar et al (24) modeled the pultrusion processing for epoxy resin with glass, carbon and keviar fibers. The model has three submodels: heat transfer, pressure and pulling force. The validity of the model was tested by comparison of the glass fiber-epoxy system with experimental values available in the literature. Different process variables were studied: wall temperature profiles, fiber content, pulling velocity and different fibers.

Model Process for Thermoset Resins

The application of this model allowed a better understanding of the relationships between processing conditions and material behavior leading to the determination of a processing window or envelop.

The pultrusion die consists of three zones. In the first zone or taper section, the resin-reinforcement mixture enters the die and there is a resin back flow which increases pressure.

In the second zone or heated die section, the resin is heated by the die wall and the chemical system starts to react and change from a viscous liquid into a gel-like material. The pressure rise is caused by resin thermal expansion. When the resin starts to be solid, the composite shrinks from the wall and it causes the pressure to decay. This effect occurs in the third die section also, which is the friction solid-solid section.

The model has been divided into three steps: heat transfer model, pressure model and pulling force model.

Heat Transfer Submodel

A cylindrical pultrusion die was considered. The following assumptions were made: i) the process is at steady state, ii) heat conduction in the axial direction is negligible, iii) there is perfect contact between composite and wall, iv) there is no heat transfer in the taper section.

The energy balance equation is:

$$\rho \, C_p V_z \frac{\partial T}{\partial z} = \frac{k}{r} \frac{\partial}{\partial r}\left(r \frac{\partial T}{\partial r} \right) + \rho_r \left(-\Delta H_{po} \right) \frac{\partial x}{\partial t} \tag{1}$$

where ρ is the composite density, C_p is the composite heat capacity and k is the composite thermal conductivity, V_z is the pulling velocity, x is the conversion, ρ_r is the resin density and $(-\Delta H_{p0})$ is the reaction heat.

Equation (1) was solved with the following initial and boundary conditions:

$$0 \leq r \leq R \quad z = 0 \quad T = T_0 \quad x = 0 \tag{2}$$

$$r = R \qquad 0 \leq z \leq L \qquad T = T_w = T_w(z) \tag{3}$$

$$r = 0 \qquad 0 \leq z \leq L \qquad \frac{\partial T}{\partial r} = 0 \tag{4}$$

where T_0 is the initial temperature of the composite at the entrance of the die.

The composite physical properties were obtained from the following equations:

$$\rho = \phi_r \rho_r + \phi_f \rho_f + \phi_g \rho_g \tag{5}$$

$$C_p = w_r C_{pr} + w_g C_{pg} + w_f C_{pf} \tag{6}$$

$$\frac{1}{k} = \frac{\phi_r}{k_r} + \frac{\phi_g}{k_g} + \frac{\phi_f}{k_f} \tag{7}$$

where r : resin, g : fiber and f: filler
ϕ: volume fraction
w: weight fraction

Pressure Submodel

The die was divided into 2 sections: a) Taper with a Backflow Model and b) Heated die with Expansion-Shrinkage Model.

Taper Section

Figure 4 shows the taper sections. Considering the system as parallel capillaries and by the application of Blake-Kozeny-Carman equation, the following equation was obtained:

$$-\frac{dP}{dz} = V_z \left(\frac{\phi_g(\xi)}{\phi_g(\xi)} - 1 \right) \frac{\eta \, \phi_g(\xi)^2}{(1 - \phi_g(\xi))^3} \frac{16K}{d_f^2} \tag{8}$$

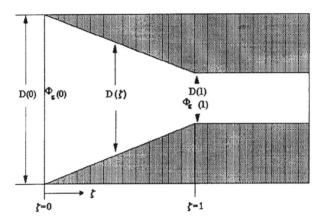

Figure 4. Scheme of taper section.

where ξ is the dimensionless length ($\xi = z/L_t$), L_t is the taper length, d_f is the fiber diameter, ($1 - \phi_g$) is the system porosity, V_z is the pulling velocity, η is the viscosity and K is the Blake-Kozeny constant.

The relation between the fiber fraction and the die position was determined by a simple mass balance:

$$\phi_g(\xi) = \phi_g(1)\, D(1)/D(\xi) \tag{9}$$

A round taper was considered and the following equation represent the die position:

$$D(\xi)/D(1) = Cr - (Cr - 1)\left[1 - (1 - \xi)^2\right]^{0.5} \tag{10}$$

where Cr is the compression ratio $= = D(0)/D(1)$. Equation (9) can be integrated by a Simpson rule method.

Heated Die Section

A state equation is needed to relate density to temperature, pressure and conversion. By the application of the constant volume condition, the following equation was found:

$$P = \frac{\alpha}{\beta}(T - T_0) - \frac{\nu}{\beta}\bar{x} + P_0 \tag{11}$$

where α is the thermal expansion coefficient, ν is the shrinkage, β is the compressibility constant and P_0 is the pressure calculated at the end of the taper.

The average temperature and conversion were calculated as:

$$\bar{x} = \frac{2}{R^2}\int_0^R x(r)\, r\, dr \tag{12}$$

$$\bar{T} = \frac{2}{R^2}\int_0^R T(r)\, r\, dr \tag{13}$$

Pulling Force Submodel

The assumption in the pulling force model is that after the taper there are two contributions to the force:

i) viscous drag (S_d) before gel
ii) friction solid-solid (S_f) from the resins to the wall and from the fibers to the wall.

$$F_{tot} = \int_0^{L_1} S_d(z)\, dA + \int_{L_1}^{L} S_f(z)\, dA \qquad (14)$$

The total pulling force was calculated by the foregoing equation: where L_1 is the position in the die where $S_d \geq S_f$.

Drag Traction

Viscous forces result from the shear at the thin layer between the fibers and the wall (λ), and these can be described by the following expression:

$$S_d = \eta \frac{V_z}{\lambda} \qquad (15)$$

The thickness of the resin layer λ can be calculated from a hexagonal array as:

$$\lambda = \frac{area}{d_f} = d_f \left(\sqrt{\frac{\pi}{2\phi_f \sqrt{3}}} - 1 \right) \qquad (16)$$

Friction Solid-Solid

The contribution of friction between the composite and the wall in the pulling force was considered as:

$$S_f = \mu P(z) \qquad (17)$$

where μ is the friction coefficient.

Influence of Processing Variables

Pultrusion model results are shown in Figure 5.

Wall Temperature Profile

The temperature profile influences the cure and viscosity behavior. The temperature activates the kinetic and chemorheology processes. The temperature profile within the oven is comprised of three parts: the initial ramp, dwell segment and final ramp. The temperature increases from room temperature until the maximum wall temperature. The initial ramp section causes decreasing viscosity of the resin close to the taper section. If the temperature close to the taper is very high, the resin will have high back flow due to the lower viscosity. In the second section, the resin will undergo an exothermic peak and the wall temperature should be lower in order to decrease the maximum composite temperature. The maximum temperature must be lower than the degradation temperature (generally 250^0C for thermoset resins). In the third section, the wall temperature should be low enough in order to cool the composite and avoid thermal residual stresses from developing.

Cure Resin Behavior

The chemical system impacts on gel conversion, and the gel point will determine any increases in modulus. For the case of the pultrusion process, the maximum pressure value is obtained at the moment when the resin layer close to the wall die is almost fully reacted. From there on, pressure begins to decrease until all the material is totally cured and with a high modulus value. Cure or gelation under pressure is desirable in the pultrusion process because it provides better surface of the composite and less void content.

Usually polyester resin shrinkage characteristics provide the best condition for gelation to occur under pressure. Polyester resin gels and continues to exhibit expansion, and suddenly shrinks at a high initial rate providing a fast pressure reduction which reduces the frictional forces proportionally.

On the other hand, the epoxy resin begins to shrink before it gels and continues to shrink at a steadily declining rate (under a condition of declining hydraulic pressure), until it is fully cured as described by Hunter (10).

Pulling Velocity

The length of each zone affects the pull load required. One major factor influencing the length of each zone is the pulling velocity. Decreasing the pulling velocity moves the gel zone and the composite separation point close to the die entrance. For a low pulling velocity, the residence time of resin in the die is longer; hence the degree of polymerization is higher

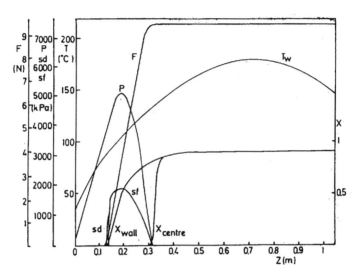

Figure 5. General behavior of the pultrusion process:
model wall temperature, T_w; center conversion, x_c;
wall conversion, x_w; pressure, P; and total pulling
force, F (See Ref.23)

and the wet-out of fiber is better and only a minor proportion of nude fiber at the wall will be found. But if the pulling velocity is very low, the production rate will be too low to be utilized in an industrial application. When the pulling velocity is very high, the die length will be not enough to achieve complete conversion into all of the composite (Figure 6)

The experimental results of Sumerak (II) for different polyester resins show that the pressure is moving down in the die heated section or increasing pulling velocities and the pulling force increases consequently (refer to Table 1 and Figure 7).

Table 1. Pulling force value as a function of pulling velocity.
(Ref.11, Sumerak's results).

System	Pulling velocity, V_z m/s x10^{-2}	Pulling Force F (KN)
31-020	0.508	0.98
	1.016	1.89
	1.524	2.18

Table 2 shows the model predictions of the pultrusion process for a Sumerak's polyester resin with low total shrinkage (1.48%). As a consequence of the superficial composite quality (24), the friction coefficient has different values. When the pulling velocity increases the friction coefficient values are close to the value of the polyester-metal friction coefficient and there is better superficial quality in the composite.

Table 2. Calculated Friccion coefficient (μ) for polyester resin with low shrinkage (Ref.1 1)

Pulling Velocity, Vz m/s x10^{-2}	μ	Pulling Force F (KN)
0.508	0.30	9.56
1.016	0.20	8.76
1.524	0.14	9.20

Fiber and Filler Content

It is known that the total volumetric shrinkage of a resin is inversely proportional to the filler or reinforcement content.

Therefore by increasing the reinforcement to resin ratio, the total volumetric shrinkage will be reduced. However, the decrease of shrinkage will produce higher total force (refer to Table 3).

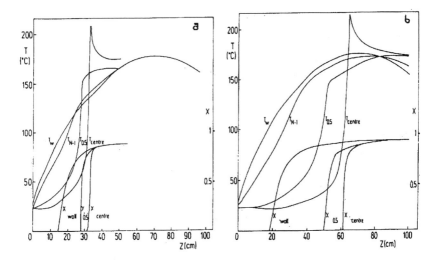

Figure 6. Temperature and conversion profiles at different radii for polyester
resin - glass fiber with a content of fiber of 0.45 (See Ref. 23)
Note: (a) $V_z=0.508 \times 10^{-2}$ m/s; (b) $V_z=1.524 \times 1^{-2}$ m/s

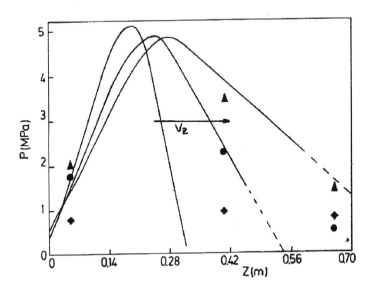

Figure 7. Pressure profiles in function of die length for differential pulling
velocities. Polyester-glass fiber system, low profile polyester.
Experimental points of Sumerak (Ref. 11) and model results
(Ref. 23). Note: Volumetric fibercontent: $Vz = 0.508$, 1.016,
and 1.524×10^{-2} m/s.

Table 3. Sumerak's result for isophtalic polyester resin and Hunter's epoxy resin for pulling velocity of 0.508×10^{-2} m/s.

System	Filler ϕ_f	Glass ϕ_g	Resin ϕ_r	Shrinkage ν	Pulling force F (kN)
31-020 [a]	0	0.50	0.50	3.96	0.98
		0.46	0.498	3.54	1.38
		0.39	0.503	2.89	6.65
Epoxy [b]	0.022	0.675	0.303	---	1.33
	0.020	0.703	0.277	---	1.78
	0.018	0.732	0.250	---	2.13
	0.016	0.747	0.237	---	3.33

[a]Ref. 11, Sumerak;   [b] Ref.10, Hunter.

An increase in fiber content decreases the maximum temperature that the material could attain, due to the diluent effect on the reaction heat. The temperature profiles for different quantities of fiber were studied for the epoxy system. The results are shown in Figure 8.

The maximum temperature decreases with the content of glass fiber and it is reached at the same place in the die. The first part of the curve shows that the system with less content of fiber has a delay of temperature because the conduction occurs from the wall to the composite center, while in the descending part of the curve, the conduction is from the center to the wall as a consequence of the chemical heat reaction.

THERMOPLASTIC PULTRUSION

State-of-the-Art in Thermoplastic Pultrusion

The composite in the form of prepreg is introduced into the die through a preheat oven or pre-heater. The taper section in thermoplastic pultrusion machine is longer than thermoset pultrusion machine. The heated die section is shorter than in the thermoset and usually it is 50% of the total die length.

Models for simulating the entire pultrusion process are primarily the same as with thermoset polymers: (a) Heat Transfer Submodel, (b)Pressure Submodel for consolidation inside the taper, (c) Pulling Force Submodel. In contrast however, there are new difficulties such as non-Newtonian matrix flow, crystallization rate and matrix melting and solidification that must be accounted for.

Lee et al. (27) developed the pultrusion process for a thermoplastic

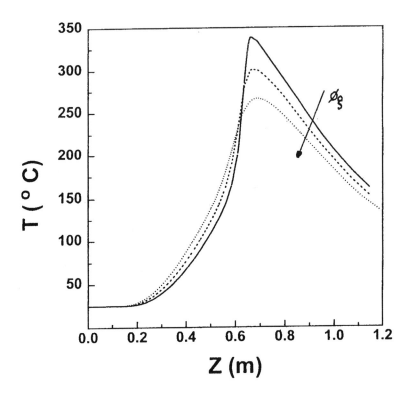

Figure 8. Center temperature profile of epoxy resin and glass fiber system with a pulling velocity of 1.016×10^{-2} m/s; and different content of fiber, ϕ_g: 0.62, 0.68 and 0.74 (See Ref. 24).

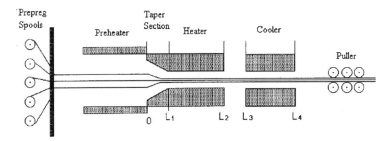

Figure 9. Scheme of Thermoplastic Pultrusion Die.

resin, APC - 2 prepreg tows. The temperature in the taper section was 427°C and the temperature in the heated die section was 149°C. The pressure distribution was calculated. As the taper angle increases the pulling force decreases. The reason is that for larger angles the length of the tapered section becomes shorter. However, for shorter dies the plies consolidate faster. After consolidation the heat transfer increases due to the increase in contact between the plies. This increase in heat transfer results in higher temperature inside the composite and lower viscosity. This decrease in viscosity further reduces the pulling force.

The heat liberated due to crystallization can be computed from the following expression:

$$Q = \left(\rho_m \phi_r H_u \frac{dc}{dt} \right) (1 - \phi_v) \tag{18}$$

where ρ_m and ϕ_r are the matrix density and matrix volume fraction, ϕ_v is the void fraction, H_u is the ultimate heat of crystallization and dc/dt is the rate of matrix crystallinity changes. The rate of crystallization dc/dt is a function of temperature and cooling rate.

Astrom and Pipes (28) showed that the importance of the thermal expansion is small (less than 2% in terms of the matrix pressure). The viscous resistance and compactation are the major contribution to the pulling force and the friction resistance is negligible. This is a consequence of the small magnitude of the pressure load reached in the die so that the matrix pressure does not contribute to the pulling force.

The investigators have shown that the pulling force increases with the increase of the pulling velocity, while the friction remains constant and low.

The same authors have a verification of the model (29) in which they show good agreement between the model and the experimental results. They worked with two systems: glass fiber and polypropylene, (glass/PP) and carbon fiber and polyether ether ketone, (carbon/PEEK). The fiber volume fraction (preimpregnated material) was 36% for the glass/PP and 55% for carbon/PEEK. Two different rectangular die configurations were studied.

In the case of carbon/PEEK system, the maximum temperature into the composite will be higher than the thermoset system (400°C). The resin and fibers were preheated from room temperature to 300°C in order to melt the resin. For the case of glass/PP the temperature in the preheated section reached 150°C. In the taper section the temperature reached 190°C . In both cases, the thermocouple location was in the superior corner of the composite cross section. The pulling velocity varied from 5 to 50 m/s. The pulling forces varied from 2.5 to 4 kN for carbon/PEEK system and it was lower for the case of glass/PP composite (200-700 N).

Viscosity is function of the temperature into the die. Lee et al (27) used an Arrhenius type function and Amstrom and Pipes (29) considered a non-Newtonian behavior of the polymer melt in order to calculate the pressure into the die for a shear-thinning fluid. The model of a non-newtonian fluid produces a relation between the pulling force and pulling velocity.

Impregnation mechanics for the melt pultrusion of thermoplastic matrix composites were described by Chandler et al. (30). A model based on lubrication theory was proposed for the impregnation of glass fiber tow when it passed over a cylindrical pin in the presence of nylon 66 resin. Different regions of behavior in pin impregnation were defined: the entry, the impregnation , the contact and the exit zone. The entry zone is the zone were the fibre tow approaches the pin. The impregnation zone is the zone where the pressure generated in the resin film causes resin to be pumped into the fibre tow. The contact zone is the region where the fibre begins the contact with the pin. In this section, friction and viscous drag results in a its high rate of build-up of tension in fiber tow. The exit zone is where the tow leaves the pin. The models application permitted the authors to optimize the resin volume fraction and the tension build-up.

CONTROL OF THE PULTRUSION PROCESS

Wu and Joseph (31) simulated the pultrusion heat transfer in the die and have determined the different conversion and temperature composite profiles for different pulling velocities. They point out the importance of automating the process by using an expert system in order to optimize the operations setpoints, manage emergency conditions and close the quality control loop with the information available, such as: temperature measurement, product quality, pulling velocity, pulling force required and the extent of reaction.

J. M. Methven and I Katramados (32) used the dielectric measurements and friction methods for monitoring a pultusion process of thermoset resin in line. The dielectric measurement was used in order to determine the resin's crosslinking.

PROPERTIES OF PULTRUDED COMPOSITES

Pultruded composites are receiving widespread applications in sporting goods, corrosion resistant parts, architecture, transportation, agriculture, chemical engineering, aircraft and aerospace due to the fact that an almost unlimited length of composites can be produced, with more flexible and higher tensile strength than those obtained in any other reactive polymer process.

Longitudinal and transverse properties are important. Continuous rovings provide longitudinal strength, while mats or resins only provide transverse strength. A pultruded part with a longitudinal strength of 965 MPa could have a transverse tensile strength of 24 MPa. For this reason, most profiles are manufactured with continuous-strand mats or other forms of transverse reinforcement. This configuration produces a laminate in which the properties are more evenly distributed (refer to Table 4).

Table 4. Properties of pultruded laminate.

PROPERTY	35-45% glass		45-55% glass	
	Parallel	Transverse	Parallel	Transverse
Tensile Strength (MPa)	207	45	310	55
Tensile Modulus (GPa)	15.8	5.5	17.2	6.9
Flexural Strength (MPa)	207	69	310	103
Flexural Modulus (GPa)	8.9	4.1	12.4	5.5
Izod Impact (J/m)	1068	214	1495	214
Compres. Strength (MPa)	138	69	207	103
Compres. Modulus (GPa)	15.8	5.5	17.2	6.9
Shear Strength (MPa)	31	31	59	59

The reinforcement are 50% mat and 50% roving by weight (33).

Thermosetting Matrices

The process is difficult to control unless one knows how to handle the exothermic chemical reactions taking place inside a pultrusion die. The highest reaction temperature must be kept below of the polymer degradation temperature (18,34). To achieve a uniform degree of cure in the cross section of a pultruded product, the temperature profile in the pultrusion die is the most important condition.

Thickness Effect

Table 5 shows the flexural properties of pultruded composites (78% w/w glass fiber -polyurethane) measured at various thicknesses (34). It can be observed that the flexural properties of composites increased with decreasing die thickness; that means, the less the die thickness, the greater the degree of conversion of composites.

Table 5. Flexural Properties for Differential Thicknesses
(78% w/w glass fiber).

MATERIAL	Die Thickness (10-2 m)	Flexural Strength (MPA)	Flexural Modulus (MPa)
GF/PU	0.319	72.6	69.20
GF/PU	0.208	80.2	7393

It has been shown through mathematical modelling (18) that for thin composites the behavior is almost isothermal and the degree of cure is uniform across the thickness. For this reason higher die temperature can be used obtaining higher reaction rates and allowing faster pulling velocities and shorter residence times. However, the presence of multiple temperature zones is necessary for thick composites in order to avoid exothermic peaks inside the die.

Effect of Pulling Velocity

Table 6 shows the effect of pulling velocity on the flexural strength of pultruded glass fiber polyurethane. The lower the pulling velocity, the longer the residence time of the resin in the die and, of course, the higher the degree of polymerization. But there is a strong compromise between the degree of polymerization (crosslinking) and degradation of the composites. Usually these kinds of composites begin to degrade at a die temperature of 180 °C. When the die temperature is lower than this, mechanical properties worsen, since the conversion is lower. The die temperature should not be lower than 170 °C; because the composites cannot be processed in a short time.

Table 6. Flexural strength as a function of pulling rate and die temperature [35],(77% w/w glass fiber-PU).

Pulling Velocity 10^{-2} m/s	Die Temperature °C			
	170	180	190	200
0.67	63	76	55	---
1	50	80	59	---
1.33	---	77	65	51
2	---	70	75	60

Effect of Filler Type and Content

Usually fillers are added to the matrix in order to reduce the cost of materials, reduce shrinkage and, sometimes, to improve the mechanical properties and superficial appearance (surfaces without fillers could become coarse and, hence, mechanical properties could decrease).

The shrinkage ratio of composites decreases with increasing filler content as can be seen in Table 7 (35).

Table 7. Flexural strength and shrinkage ratio as a function of filler content in a PU-77% w/w glass fiber pultruded composite.

Filler content (phr)	Calcium carbonate (CaCO$_3$)		Mica	
	shrinkage (%)	σ_b (MPa)	shrinkage (%)	σ_b (MPa)
0	9.8	75	9.8	75
10	3.3	83	4.9	83
20	2	85	3.5	86
30	1.2	91	2.3	88
40	0.4	105	1.5	90

Effect of Fiber Type and Content

Experimental results indicate that tensile strength, notched izod impact strength, flexural strength and flexural modulus increase linearly with fiber content, no matter which type of fiber is used (35). This behavior is characteristic for systems with continuous and aligned fibers, where fiber mechanical properties are superior in comparison with the matrix.

Thermoplastic Matrices

The major limitation to wide-spread use of thermoplastic resin for pultrusion is their poor fiber wet out due to high melt viscosity (500 - 1000 Pa.s for thermoplastics versus 0.1 - 2 Pa.s for thermosets).

Since pultrusion is an automatic and continuous process for fabricating fiber reinforced plastics, a proper pulling velocity will optimize production rate and mechanical properties.

Processing Variables

Table 8 shows the effect of pulling velocity on the tensile, flexural and impact strength of pultruded glass fiber (75% w/w) reinforced poly (methyl metacrylate) composites (36).

It should be noted that the methyl methacrylate prepolymer is first synthesized from its monomer and then used directly and polymerized in the die (in situ thermoplastic pultrusion concept).

Table 8. Mechanical properties for different pulling velocities (PMMA-glass fiber).

Pulling velocity 10^{-2} (m/s)	Tensile Strength (MPa)	Flexural Strength (MPa)	Notched Izod Impact Strength (KJ/m)
0.66	720	415	2.4
1.17	500	200	1.0
1.67	415	138	0.64

In general, mechanical properties increase with decreasing pulling velocity. For a lower pulling velocity, the residence time of the resin in the die will be longer; hence the degree of polymerization will be higher. But for too low a pulling velocity, the production rate will be too low to be utilized in industrial applications.

Die Temperature Effect

The higher the temperature, the faster the polymerization rate and the lower the degree of polymerization. For PMMA, the number and weight average molecular weight increases with decreasing die temperature, but it must be high enough to process the composites in a short time. At very high temperatures the molecular weight will decrease sharply due to degradation.

Table 9 shows that the lower the die temperature the better the mechanical properties. When the pulling velocity of glass fiber reinforced PMMA composites is less than 0.01 m/s, the mechanical properties increase with decreasing die temperature. But when the pulling velocity is above 0.01 m/s and a low temperature, the monomers do not polymerize completely and the mechanical properties decrease sharply.

Table 9. Mechanical properties vs die temperature and pulling velocity.

Die Temp. (°C)	Tensile Strength (MPa)			Flexural Strength (MPa)		
	Pulling Velocity 10^{-2}(m/s)			Pulling Velocity 10^{-2}(m/s)		
	0.67	1.17	1.67	0.67	1.17	1.67
140	720	500	415	415	200	138
160	552	520	500	280	218	192
180	500	467	448	192	145	135

Glass fiber-PU pultruded parts show a similar behavior of their properties with die temperature. But in all cases a compromise between the die temperature and the pulling velocity is needed, otherwise serious degradation problems appear.

COMPARISONS BETWEEN PULTRUDED MATERIALS

Tables 10 and 11 show pultruded materials with glass fiber and carbon fiber respectively. For both tables the following abbreviations have been used: PP-polypropylene, PMMA-poly (methyl methacrylate), PU-polyurethane, NY6-nylon 6, PPS-polyphenylene sulfide, UP-unsaturated polyester, PH-phenolic resin and ABS-acrylobutadiene styrene.

Tables 10. Properties of Glass Fiber Pultrused Composites.

Properties	PP	PMMA	PU	NY6	PPS	UP	PH	ABS
Fiber content (wt %)	58	75	75	75	72	75	73	75
Tensile strength (MPa)	---	848	890	869	793	828	448	710
Flexural strength (MPa)	451	579	72	469	965	828	483	538
Notched Izod impact (KJ/m)	---	2.0	unbrok.	2.4	3.1	2.1	2.1	2.5
Flexural Modulus (GPa)	31	---	---	---	---	---	---	---
Shear strength (MPa)	23	---	---	---	---	---	---	---
References	43	36	37	41	39	39	42	40

Tables 11. Properties of Carbon Fiber Pultruded Composites.

Properties	PMMA	PU	NY6	PPS	PH	EPOXY
Fiber content (wt %)	58	58	57	56	58	54.4
Tensile strength (MPa)	1503	1525	1496	1172	1103	1917
Flexural strength (MPa)	434	86	498	1365	827	1213
Notched Izod impact (KJ/m)	1174	unbrok.	1708	1601	2562[a]	2364[a]
References	36	37	41	39	42	38

[a]Notched Charpy Impact Strength.

REFERENCES

1. W. V. Breittgam and W. P. Ubrich, Revista de Plasticos Modernos. 417, 374-382, (1991)
2. J. Martin and J.E. Sumerak, Pultrusion, Engineered Materials Handbook-Composites, voll, ASM International, 533-543,1987
3. J.K. Rogers, Revista de Plasticos Modernos. 416, 257-260, (1991)
4. Modern Plastics International, April, 73,(1989)
5. N. Fishman, Modern Plastic International, August, 50-51, (1989)
6. B.T. Astrom and R.B. Pipes, Polymer Composites. 14.3. 184-194, (1993)
7. J. E. O'Connor and W.H. Beever, Revista de Plasticos Modernos. 383,

756-758, (1988)
8. B. W. Ewald, Modern Plastics Encyclopedia. Mc.Graw-Hill, 318, (1985-86)
9. N. J. Tessier, D. Kiernan, A. Madenjian and G. Moulder, Revista de Plasticos Modernos. 371, 647-650,(1987)
10. G. A. Hunter; 43rd Ann. Conf. RP/C,SPI, 6-C, 1988.
11. J.E. Sumerak, Revista de Plasticos Modernos. 356,feb.(1986)
12. L.J.Lee, Makromol.Chem..Macrom.Svmp..68.169/1993)
13. J.M.Kenny and M. Opalicki, Makrom. Che., Macromol.Symp., 68, 41-46, (1993)
14. Y.A. Tajima and D.C. Crozier, Poly. Eng. and Sci. 28, 7 (1988)
15. C.D.Han, D.S.Lee and H.B. Chin, Polym. Eng. Sci. 26. 6, 393(1986).
16. C.D.Han and H.B Chin, Polymer Eng. Sc., 28, 5, 321, (1988)
17. H. NG and I.Manas-ZIoczower, Polym. Eng. Sci., 29.1097(1989)
18. A. Trivisano, A. Maffezzoli, J.M.Kennyand L.Nicolais, Adv. in Polymer Technology. 10,4, 251-264, (1990).
19. G.L. Batch, PhD Thesis, Minnesota University, (1989)
20. C. Vallo and A. Vazquez, Eng.Plastics. 5,2,77, (1992)
21. Price H.L., PhD Thesis, Old Dominion University, (1979)
22. G.L. Batch, Revista de Plasticos Modernos, 419 may.(1991)
23. S.M.Moschiar, M.M.Reboredo, J.M.Kenny, and A.Vazquez; Polymer Composites, in press.
24. S.M.Moschiar, H. Larrondo, M.M.Reboredo and A.Vazquez; Polymer Composites, accepted.
25. Bibbo, M.A. and Gutowski, T.G..Technical Paper. Soc.Plast.Eng., 32, 1430, (1986)
26. H-T Wu and B. Joseph, SAMPE Journal , 26, 6, (1990)
27. W. 1. Lee, G.S. Springer and F.N. Smith, J.of Composite Materials., 25, 1632, (1991)
28. B. T. Astrom and R.B. Pipes, Polymer Composites, 14, 3, 1184-194, (1992)
29. B. T. Astrom and R.B. Pipes, Polymer Composites, 14, 3, (1992)
30. H.T. Wu and B.Joseph, SAMPE Journal, 26, 6, (1990)
31. H. W. Chandler, B.J. Deviin and A.G. Gibson, Proceeding of the Ninth International Conference Composite Materials, (ICCM/S), Vol. III, 737-744, (1993)
32. J.M.Methven and I Katramados, Proceeding Polymat'94 Conference, pp 321-324
33. D. Evans; "Pultrusion" in Encyclopedia of Polym Sc. Eng., John Wiley and Sons, 1988, Vol 13, Pg 660.
34. Ma,C.M.; Chen,C.; J.App.Polym.Sc., 50, 759 (1993).
35. C. Chen. C. Ma: J.Amy.Polym.Sc..46. 949 (1992).
36. C.M. Ma, C. Chen, J.App.Polym.Sc., 44, 807(1992).
37. C. Chen, C. Ma, J.App.Polym.Sc., 46, 937 (1992).
38. I.S. Hwang; Master Thesis, National Tsing Hua University, Taiwan, 1986.
39. J.E. O'Connor, W.H. Beever. 42nd Ann. Conf.. RP/C, SPI, I-D, 1987.

40. J.S. Hwang , S.N. Tong , S.J. Tas, P.H.H. Hsu and P.T.K. Wu; 43rd Ann. Conf.. RP/C, SPI, 6-E, 1988.
41. C. Ma, M.S. Yn; Taiwan, R.O.C. Pat 35969, 1989.
42. C. Ma and W.C. Shih,W.C.; U.S. Pat 4873128, 1989.
43. W.J. Tomlinson and J.R. Holland, J.Mat. Sci. Let.. 13, 675(1994).

7

Electron Beam Processing of Polymers

Sujit K. Datta, Tapan K. Chaki and Anil K. Bhowmick
Rubber Technology Center, Indian Institute of Technology
Kharagpur, INDIA

INTRODUCTION

Radiation processing of polymers was introduced after world war II with the development of the nuclear reactor. Basic research, started in the late 1940's are continuing with the commercialization of the technology [1]. In recent years various radiation sources e.g., X-rays (soft and hard), gamma (γ) and ultraviolet (UV) rays and electron beam (EB) are being used in the curing of various polymers, paints, ink, coatings, adhesives etc and in graft copolymerization. Radiation processing particularly with electron beam offers some distinct advantages, some of which are:

1. The process is very fast, clean and can be controlled precisely.

2. There is no permanent radioactivity since the machine can be switched off.

3. The electron beam can be steered very easily to meet the requirements of various geometrical shapes of products to be irradiated. This cannot be achieved by X-rays and γ-rays.

4. The high penetrating power of the electron beam allows the efficient curing of thick polymeric articles, highly pigmented inks and coatings. Pigmented inks and coatings cannot be otherwise cured by UV radiation due to low penetrating power.

5. The EB radiation process is practically free of waste products and hence there is no serious environmental hazard.

6. It gives large throughputs compared to other radiation processes.

7. It can be easily implemented at any stage of production.

Now-a-days electron beam radiation processing has wide applications particularly in the wire, cable, coating and the tire industries. This chapter deals with the processing of polymers by electron beam radiation and the properties as well as the applications of modified polymers.

WHAT IS ELECTRON BEAM PROCESSING ?

Like infrared and microwave, electron beam is an electromagnetic radiation except that its wave length is very short and the frequency is very high. Table 1 gives comparative chart of various electromagnetic radiation, EB has greater penetrating power as compared to other radiation processes and is associated with very high energy. However, EB is classified as ionizing radiation . This is different from the microwave and the infrared radiation which are nonionizing.

Table 1. Frequency and wave length of various radiations.

Radiation	Frequency (Hz)	Wavelength (μm)
Infrared	10^{15}-10^{12}	1-10^2
Ultraviolet	10^{17}-10^{15}	10^2-1
Microwave	10^{12}-10^{10}	10^3-10^5
Electron beam	10^{21}-10^{18}	10^{-7}-10^{-4}

In the electron beam process, the electrons are generated by the electron gun or injector by thermo-ionic emission and accelerated through a potential of 150 to 250 thousand volts under vacuum of 10^{-6} torr. These are then passed through the window into the target. The machine in which the electrons are generated and accelerated is conventionally termed an electron accelerator or EB generator. Figure 1 shows the schematic diagram of a typical electron beam accelerator.

The main part of an accelerator is obviously the electron injector system or electron gun placed on the upper portion of a cavity called the resonator cavity. The electron injector produces a triode comprising the cathode (generally made of LaB_6), anode and a control grid. An equivalent electron gun circuit is shown in Figure 1. A conical spiral heater made of tungsten wire of about 0.6 mm in diameter is attached with the cathode for heating. The electrons produced in the cathode by thermo-ionic emission, are accelerated through the accelerating gap by a high voltage. The accelerated beam current is controlled by changing the value of the positive bias voltage on the cathode with respect to the grid.

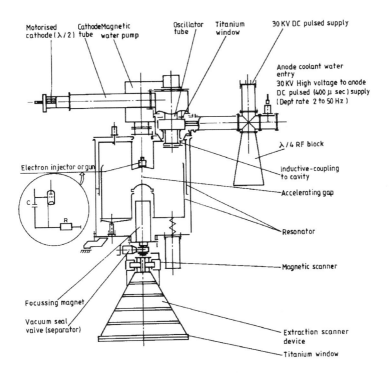

ELECTRON ACCELERATOR ASSEMBLY

Figure 1. Schematic diagram of typical Electron Beam Accelerator
and equivalent circuit diagram of electron gun.

The energy of accelerated electrons is around 0.5 to 2.5 MeV and the beam power is about 20 kW over the entire energy range. The unit of radiation dose is the Gray (Gy) and is defined as the dose required by 1 kilogram of material to absorb I Joule of energy.

1 Gy = 1 J in 1 kg = 1 w-sec in 1 kg = 100 Rad = 10^{-4} Mrad

The dose rate (D_R) for an electron accelerator can be written in terms of beam current (I) and irradiation field area (R)

$$D_R = K\ I/R$$

where K is a stopping power of electrons, which depends on the energy of electrons and density of materials being irradiated [2]. The current can be varied between extremely low to high values which provide good flexibility for the utilization of the accelerator for processing applications.

A magnetic lens is placed just after the anode to control the transverse dimension of the beam at the entrance into the extraction device. The beam extraction device is a fan shaped body of stainless steel (about 1 mm thick) located below the electron gun. The electron beam emerges through a titanium foil of 50 micron thickness . The efficiency and quality of radiation products mainly depends on the methods of beam extraction from the vacuum accelerating chamber into the atmosphere. Different types of extraction devices are used depending on the geometry of products and type of irradiation e.g. linear scanning extraction device for single sided irradiation, fir-like scanning device for double sided irradiation, quasiring scanning for three sided irradiation, etc. [2].

An important point should be noted here that some ozone is formed during irradiation which is harmful to polymers and should be removed. Typical ozone output within an air gap between an irradiated subject and an extraction foil of 0.05 mm thickness is 60 mg/sec.

GENERAL EFFECTS OF ELECTRON BEAM ON POLYMERS

General effects of electron beams on polymers have been recently discussed by Bhowmick [3]. Since chemical bonds only require a few ev for destruction, it might be expected that the incident radiation of an electron beam of typically several million ev would cause bond destruction at random. In fact, this is not so. The reactions involved are remarkably selective [4]. The reactions are primary and secondary in nature. Primary reactions are two folds : Ionization, the production of an ion and excitation, the formation of fast electrons of excited chemical species [5].

A $\Longrightarrow$ $(A^+)^*$ + e^- (Ionization)
A $\Longrightarrow$ A + e^- (Excitation)

A $\Longrightarrow$ A*   (Excitation)

In the secondary reactions, the excited ions dissociate into a radical ion and a radical.

$(A^+)^* \Longrightarrow B^{(+)} + C$

With further reactions, excited molecules are formed. These may dissociate into secondary ions, molecular products and free radicals. The polymers may then undergo chain scission, crosslinking and recombination of broken chains.

Polymers typically undergo simultaneous scission and crosslinking but in most cases one or the other clearly predominates. The structure of polymers decides often the preponderance of one process over another. For example, polystyrene where an hydrogen atom of the polyethylene repeating unit is replaced by a phenyl ring shows a crosslinking reaction. On the other hand, main chain scission occurs and molecular fragments are liberated for polymethyl methacrylate, as branching induces weak points in the molecular structure. Table 2 summarizes the behavior of some common polymers towards crosslinking and degradation. Dominant process is degradation when a carbon is linked to 4 other carbon atoms, whereas crosslinking occurs when such a carbon is linked to at least one hydrogen atom [6].

Table 2. Classification of polymers according to the predominating behavior on irradiation.

Degradation	Crosslinking
Polyisobutylene	Polyethylene
Poly(α-methyl styrene)	Polypropylene
Poly(vinylidene chloride)	Poly styrene
Poly(vinyl fluoride)	Poly(vinylchloride)
Poly(chlorotrifluoro ethylene)	Poly(vinyl acetate)
Poly(methyl methacrylate)	Poly(vinyl alcohol)
Poly(vinylbutyral)	Poly(vinylmethylether)
Poly(acrylo nitrile)	Poly(butadiene)
Poly(methacryl amide)	Poly(chloroprene)
	Poly(ethylene vinyl acetate)
	Polyesters
	Polyurethanes
	Polyacrylates

The sensitivities of different polymers towards crosslinking and chain scission can be compared from their G values which describe the yield of radiation induced reactions. The G value is the number of molecules or ions produced or destroyed per 100 ev energy absorbed. For crosslinking and chain scission, the G value is written as G(X) and G(S) respectively. For polyethylene the crosslinking yield [G(X)] is 1.0-2.5; whereas for polystyrene it is 0.035-0.50. For polyisobutylene and for poly (α-methyl styrene) the G values for scission [G(S)] are 5 and 0.25 respectively. The presence of an aromatic ring in the polymer exerts a strongly stabilizing influence on the yield of radiation induced crosslinking or scission, and polymers bearing aromatic functionality are particularly useful in applications where radiation resistance is required [7].

USE OF MULTIFUNCTIONAL MONOMERS

It is clear from earlier discussions that some polymers degrade to a considerable extent during EB irradiation. The degradation or chain scission is greater at a higher radiation dose due to absorption of higher energy. In order to utilize the EB technology efficiently, it is mandatory that the degradation reactions must be prevented or the desired reactions must be completed within a short time before any substantial level of breakage takes place. The present day technique is to adopt the later process by incorporating multifunctional monomers shown in Table 3. These monomers produce a larger number of radicals at a much faster rate than the parent polymers so that the desired reactions are forced at a low dosage of irradiation. The multifunctional monomers are called radiation sensitizers. These monomers are also effective plasticizers. The mechanism of functioning of these monomers will be discussed in a later section.

The rate of crosslinking can also be accelerated by using halogenated compounds, nitrogen oxide, sulfur monochloride, maleimides, thiols, acrylic and allyl compounds [8]. It may be noted here that the degradation reactions may be reduced by chemical modification of polymers. For example, butyl rubber is degraded by ionizing radiation. Regular butyl rubber is a copolymer of isobutylene and a small amount of isoprene (less than 3 wt%). When the reactivity of the isoprene unit is increased by halogenation, the crosslinking reactions dominate until the active sites are exhausted. The fact that crosslinking is dominant in the early stages of the processing is indicated by increases in the Mooney viscosity and green strength of the halobutyl rubber.

ANTIRAD COMPOUNDS

Antirad compounds decrease the rate of reactions including both crosslinking and chain scission. As a result they often provide protection against polymer degradation. Antirads include aromatic amines, quinones, aromatic sulfur and nitrogen compounds. These materials are highly

Diethylene glycol dimethacrylate:

$$\underset{\text{M.W 242}}{CH_2=\overset{\overset{\displaystyle CH_3}{|}}{C}\text{-}COO\text{-}R\text{-}OC\text{-}\overset{\overset{\displaystyle CH_3}{|}}{C}=CH_2:}$$

R = CH_2CH_2O,

Trimethylolpropane trimethacrylate:

$$\underset{\text{M.W 338}}{CH_3CH_2C(CH_2OO\overset{\overset{\displaystyle CH_3}{|}}{C}=CH_2)_3}$$

Triallyl cyanurate

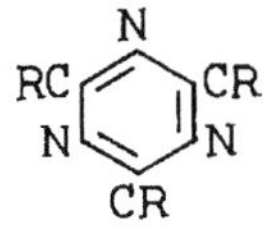

M.W 249

R = $CH_2=CH\ CH_2O$

Tetramethylol methane tetraacrylate:

$C(CH_2OOCH=CH_2)_4$
M.W 352

Hexamethylene diisocyanate:

$$(CH_2)_6(NH\overset{\overset{\displaystyle O}{\|}}{C}OCH(R)_2)_2$$

M.W 624

$$R = CH_2OCO\overset{\overset{\displaystyle}{|}}{C}=CH_2$$
$$CH_3$$

Toluene diisocyanate:

M.W 770

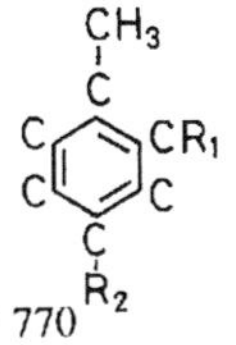

$R_1 = NHCOOC(R_2)_3$

$$R_2 = CH_2OC\overset{\overset{\displaystyle O}{\|}}{C}CH=CH_2$$

Dipenta erithritol acrylate:

$(R)_3\text{-}CCH_2OCH_2C\text{-}(R)_3$
M.W 500

R= $CH_2=COOCH_2$

Table 3. List of monomers. These monomers produce high yields of radicals during irradiation. They are also effective plasticizers.

conjugated and decrease the energy of the electrons coming out of the accelerator, thus diminishing their potential to cause polymer degradation [9].

MODIFICATION OF POLYMERS BY ELECTRON BEAM

Preparation of Tile Compounds

The plastic or rubber samples are mixed with the sensitizer at the desired level in a Brabender Plasticorder or an internal mixer. The speed and the temperature of mixing depend on the nature of polymer. A temperature of 120°C may be used for polyethylene and ethylene vinyl acetate copolymer. A higher temperature is needed for crystalline plastics.

These compounds may be molded under compression to prepare sheets of various dimensions. The compounded sample may be extruded in a single screw extruder to make tubes or cables. The inner liner compounds in the industry may be calendered first to a thickness of 1-2 mm before irradiation.

Irradiation of Samples

The articles of desired shape are subjected to electron beam irradiation at various doses (20, 50, 100, 150, 200 kGy etc.) at room temperature. Figure 2 shows the arrangement for irradiation of polymer sheet and cable or tubes.

The extent as well as the uniformity of curing depends on the penetration of energized electrons. The extent of penetration in turn depends on the density and thickness of the material to be irradiated. In general, single sided radiation is suggested for the articles of 6 mm thickness having equivalent density to water. But the materials having a higher density than water are recommended for double sided radiation. The thickness of the articles for uniform penetration can be stipulated as [10]:

$$T\,(mm) = \frac{8.0 \times E\,(MeV)}{P\,(gm/cc)}$$

where: P = material density
          E = electron energy
          T = thickness

Mechanism of Crosslinking and Grafting Reactions

The mechanisms of crosslinking and grafting reactions are studied with the help of infrared spectroscopy, photoelectron spectroscopy, reflection

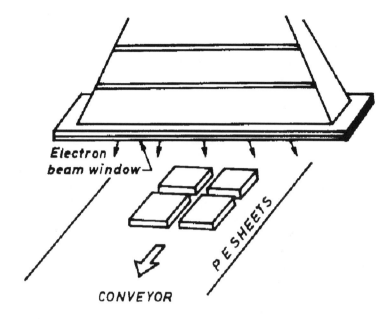

IRRADIATION OF PE SHEETS

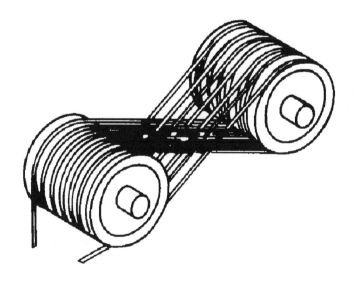

WIRE CONVEYOR FOR EB IRRADIATION

Figure 2. Arrangement for irradiation of polymer sheet and cable or tube.

energy loss spectroscopy etc. The extent and nature of the reaction depends on the structure of the polymer and the sensitizer. As there are a large number of polymers which can be processed with the help of electron beam, it is not possible to cover all the mechanisms in this chapter. However, an attempt has been made here to explain the grafting and crosslinking reactions of polyethylene and ethylene vinyl acetate (EVA) copolymer mixed with trimethylol propane trimethacrylate (TMPTMA) or trially cyanurate (TAC) in the presence of electron beams.

From the IR absorbance spectra (Figure 3) an unirradiated TMPTMA based sample shows a sharp peak at around 1640 cm^{-1} which disappears upon irradiation with 2 Mrad (20 kGy) of dose. In the case of a TAC based sample, a new peak at around 1565 cm^{-1} appears and remains unchanged on irradiation. There is no trace of sensitizers (TMPTMA/TAC) found in ether of the extracts of irradiated EVA-TMPTMA/TAC samples. The disappearance of the peak at 1640 cm^{-1} is probably due to the consumption of transvinylene group [11] by grafting and crosslinking of unsaturated acrylates (TMPTMA) onto EVA [12,13]. The appearance of the new peak at 1565 cm^{-1} in the spectra of the TAC-EVA based system is due to the presence of cyclic $>C=N$ groups present in TAC [11]. These remain unchanged during grafting of TAC onto EVA [14]. Scheme-1 in Figure 4 describes the mechanisms of grafting of TAC onto EVA followed by crosslinking. It has been reported earlier that polyethylene reacts with the multifunctional monomer in a similar manner. The residual unsaturation in some of the PVC films was very high inspite of all the monomer being consumed in the early stages of irradiation. This increases with the TMPTMA concentration and decreases with irradiation dose. As the irradiated control sample did not show any development of unsaturation, all the residual unsaturations observed in the crosslinked material must almost certainly be due to unreacted double bonds from TMPTMA [17].

The detailed mechanisms of reactions of EVA with TMPTMA and polyethylene with TMPTMA or TAC are similar to that shown in Scheme-1 and have been described earlier [12-16]. Figure 5 shows the variation of grafting level of TAC onto EVA with irradiation dose and TAC level. The grafting level does not change over the whole range of radiation dose because the grafting of TAC is almost completed at the lower radiation dose. The grafting level increases with the concentration of TAC.

During the irradiation of polyethylene and EVA, some carbonyl groups are formed due to aerial oxidation [12,15]. The absorbance due to $>C=0$ stretching at 1737 cm^{-1}, increases with irradiation dose (Figure 6) due to availability of larger numbers of free radicals for reactions with oxygen (refer to Figure 7 - Scheme-2).

At a higher irradiation dose a number of free radical initiated reactions like chain scission, disproportionation, recombination of macro-radicals take place (refer to Figure 8 - Scheme-3). With the increase of vinyl

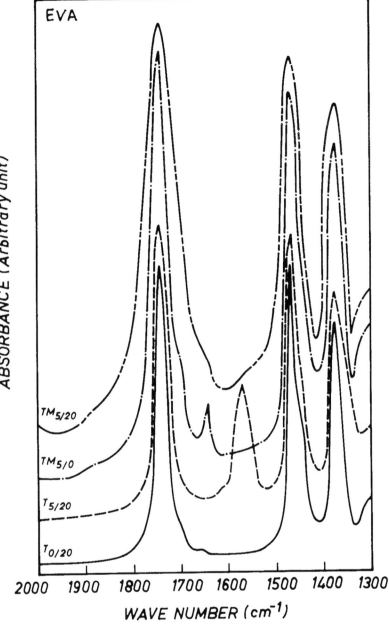

Figure 3. IR absorbance spectra of irradiated EVA ($T_{0/50}$),
unirradiated EVA/TMPTMA blend ($T_{5/0}$),
unirradiated EVA/TMPTMA blend ($T_{5/20}$),
unirradiated EVA/TAC blend ($T_{5/20}$),
[subscript indicates the polyfunctional monomer level (TMPTMA/TAC)
in percent and radiation dose in kGy respectively].

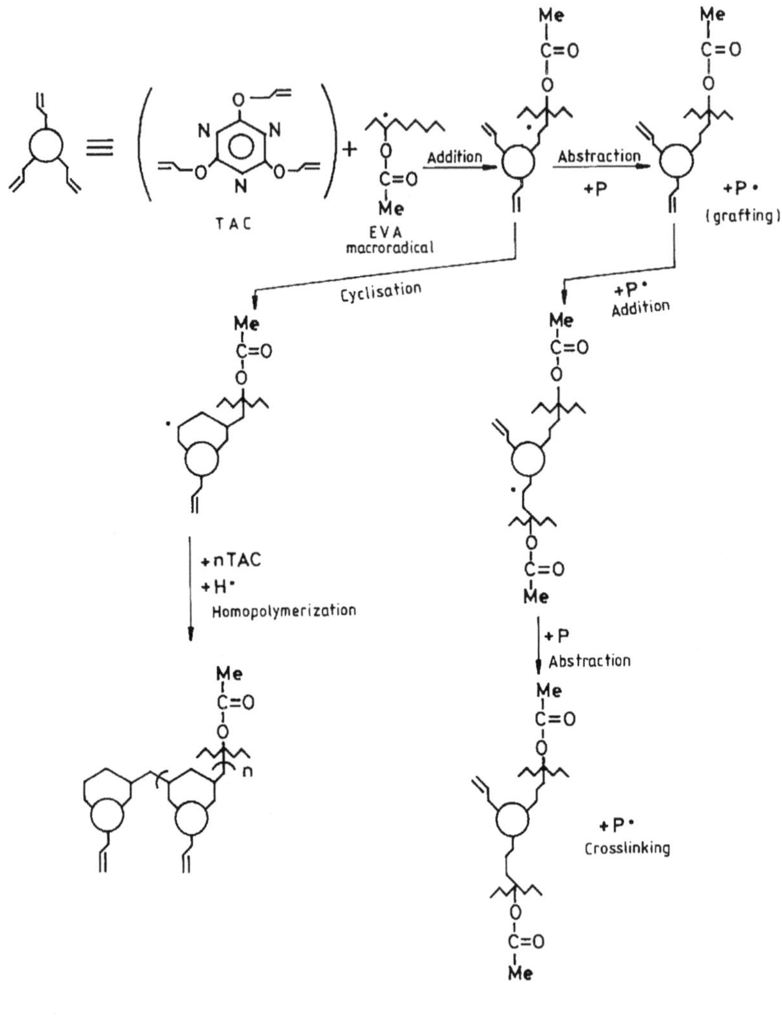

Figure 4. Illustrates the Scheme 1 reaction.

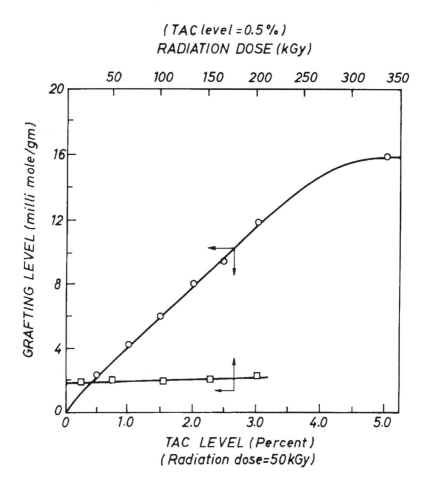

Figure 5. Plot for grafting level vs. radiation dose and TAC level.

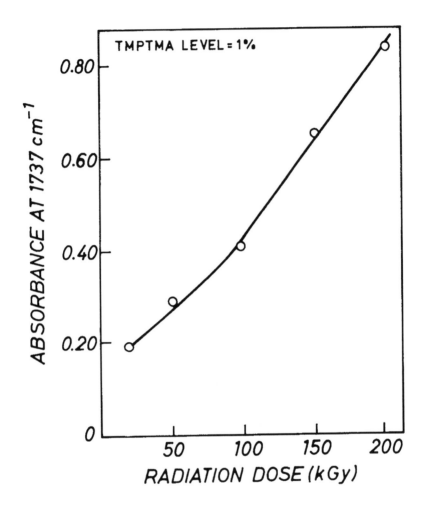

Figure 6. Variation of absorbance at 1737 cm-1 (due to $>$C=0 stretching) with radiation dose.

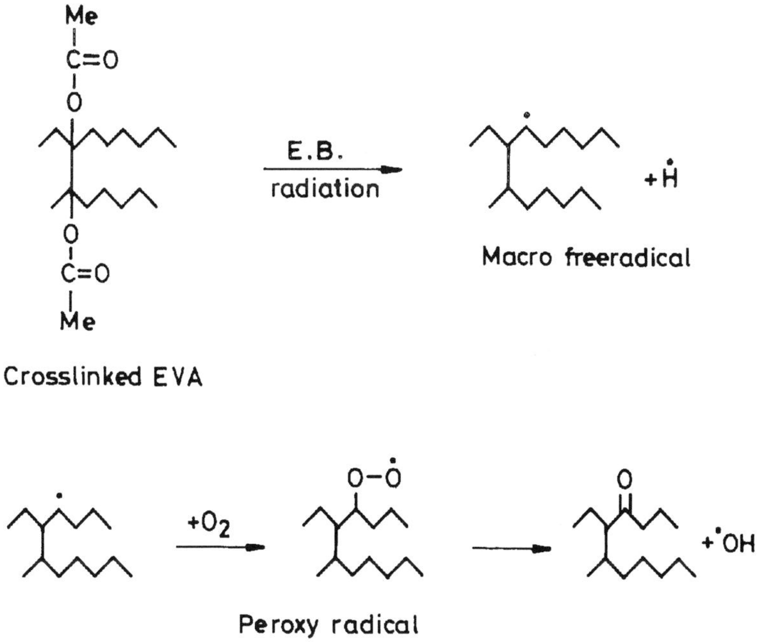

Figure 7. Illustrates the Scheme 2 reaction.

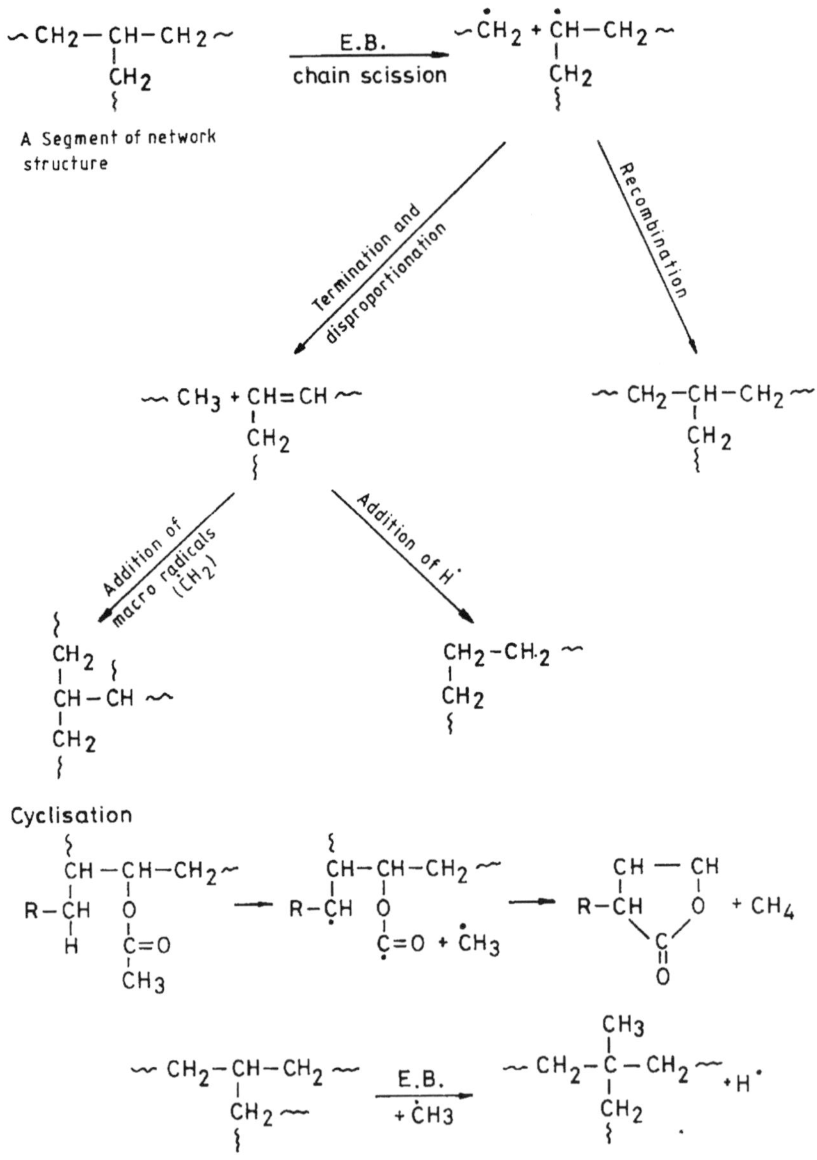

$\sim CH_2 - CH - CH_2 \sim$
       |
       CH_2
       $\{$

$\xrightarrow[\text{chain scission}]{\text{E.B.}}$

$\sim \overset{\bullet}{C}H_2 + \overset{\bullet}{C}H - CH_2 \sim$
             |
             CH_2
             $\{$

A Segment of network
structure

Termination and disproportionation

Recombination

$\sim CH_3 + CH = CH \sim$
           |
           CH_2
           $\{$

$\sim CH_2 - CH - CH_2 \sim$
            |
            CH_2
            $\{$

Addition of macro radicals ($\dot{C}H_2$)

Addition of H·

$\{$
CH_2
|
$CH - CH \sim$
|
CH_2
$\{$

$CH_2 - CH_2 \sim$
|
CH_2
$\{$

Cyclisation

$\{$
$CH - CH - CH_2 \sim$
$R - CH$ O
|
H $C = O$
     |
     CH_3

$\longrightarrow$

$\{$
$CH - CH - CH_2 \sim$
$R - \overset{\bullet}{C}H$ O
       $\overset{\bullet}{C} = O + \dot{C}H_3$

$\longrightarrow$

$CH - CH$
$R - CH$ O $+ CH_4$
      $\underset{O}{\overset{\parallel}{C}}$

$\sim CH_2 - CH - CH_2 \sim$
           |
           $CH_2 \sim$
           $\{$

$\xrightarrow[+ \dot{C}H_3]{\text{E.B.}}$

          CH_3
          |
$\sim CH_2 - \overset{}{C} - CH_2 \sim$ $+ H\cdot$
         |
         CH_2
         $\{$

Figure 8. Illustrates the Scheme 3 reaction.

of the same composite sample. Laboratory test results based on a hot xylene extraction for 18 hours, however, did not leave any solid residue. Thus solvent extraction test result failed to support the reason for an increase in the strength of very lightly filled PP composite. Nevertheless, it can be argued that some minor chemical interaction did occur during the mixing and molding processes which were not strong enough to withstand the extreme thermal condition imposed during solvent extraction process.

A difference in the absolute value of the ultimate properties was observed if the type of filler component was changed. The tensile strength of zeolite-filled composite demonstrates a   maximum   for the highest concentrations of BMI and zeolite in the chosen experimental range. This fact indicates that degree of interaction between PP and zeolite is directly proportional to the concentration of filler for a BMI concentration of about 2.5 wt%.

Table 5. Impact strength of filled composites.

Filler type	wt % filler	Reactive modifier	Impact strength, $kJ.m^{-2}$
$CaCO_3$	40	none	4.2
	40	BMI	9.7
Zeolite	40	none	4.9
	40	BMI	7.2
$Mg(OH)_2$	60	none	2.4
	60	BMI	9.3
Talc	40	none	8.1
	40	BMI	9.1

The impact strength of the composite has a negative influence on the filler concentration. In general, the more the loading, the less the resistance to impact at low temperature. The absolute value of the impact strength, however, has a marginal positive influence on the BMI concentration compare to that of the unmodified ones [21,22]. Table 5 furnishes the impact strength values of several composites at low temperature (-20^0C). The improved impact strength of reactively modified composites can be attributed to better stress distribution either due to the changes in the phase crystallinity and/or due to the formation of chemical linkages.

The impact strength of a thermoplastic matrix is dependent on the molecular characteristics including polymer chain length and degree of branching. Usually, a minor interaction between chain molecules might help to facilitate the energy transfer due to the application of sudden acetate content the absorbance ratio of 1737 cm^{-1} to 1470 cm^1 increases (Figure 9). As the vinyl acetate content increases, the mobility of the α-hydrogen atoms (with respect to -OCOCH$_3$ group) increases. As a result, these split off easily upon irradiation and the number of macro radicals increases with the vinyl acetate content. These radicals then react with oxygen (Scheme-2, Figure 7).

PROPERTIES OF MODIFIED POLYMERS

Gel Content

Crosslinking takes place upon subjecting the EVA or PE compounds to irradiation. As a result the gel content is expected to increase with an increase in radiation dose. But in practice the gel content increases initially with radiation doses up to 100 kGy for EVA and 50 kGy for PE after which the change becomes insignificant (refer to Figure 10). This is because crosslinked structure breaks down at higher irradiation doses. With the increase in the concentration of multifunctional monomer (TMPTMA or TAC) level the gel content remains almost constant as in the case of EVA. Due to the balancing effect between crosslinking and chain scission taking place simultaneously, the gel content does not change significantly with the increase in multifunctional level. However the gel content increases linearly with the vinyl acetate content in EVA (Figure 9) at a particular radiation dose (50 kGy) and sensitizer level (1%). As the vinyl acetate content increases, the number of most probable crosslinking sites is increased and the crosslinking takes place to a greater extent [18].

Tensile Properties

Tensile strength and elongation at break of EVA with 1 % TMPTMA increase with the radiation dose up to 50 kGy. On further increase of the dose there is a decrease of these properties (Figure 11). This may be due to the fact that up to 50 kGy radiation dose and 1 % TMPTMA level large network structure is formed. At higher radiation dose or polyfunctional monomer level these network structures begin to breakdown. PE samples show a similar behavior, although the maximum at which the peak appears is different. For example, the plot of tensile strength versus radiation dose indicates a peak at 100 kGy dose for PE and 50 kGy for EVA. The optimum radiation dose decreases with the increase in vinyl acetate content (Figure 12). These are possibly related to the radical generation, and to the participation of these radicals in the crosslinking and chain scission. Chain scission is more predominant in the case of EVA. The tensile strength, as shown in the Figure 13 is related to the energy to break - the higher the energy to break, the higher the tensile strength. Modulus at 100% elongation increases marginally with radiation dose or monomer level due to crosslinking. A similar type of observations is made with other polymer-multifunctional monomer combinations [12-19].

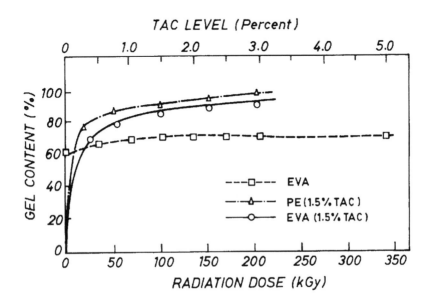

Figure 10. Variation of gel content of EVA and PE compounds with radiation dose and TMPTMA level.

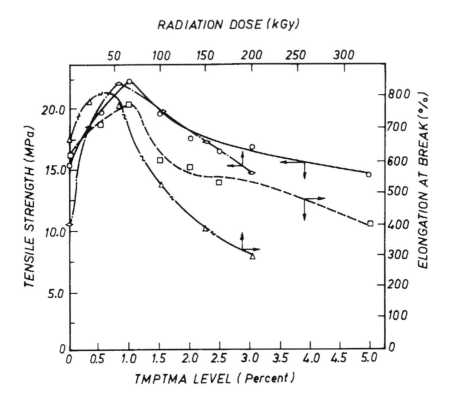

Figure 11. Plots of tensile strength and elongation at break with radiation dose and TMPTMA level.

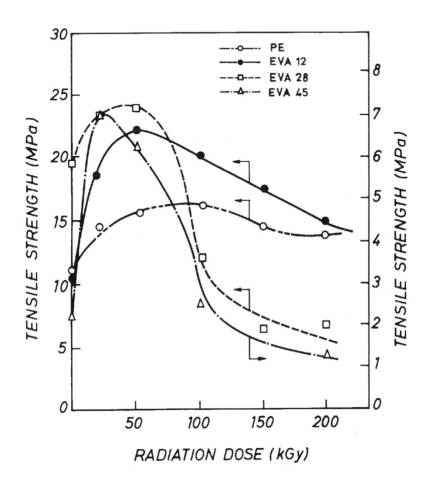

Figure 12. Plots of tensile strength vs. radiation dose of PE, EVA 12, EVA 28 and EVA 45.

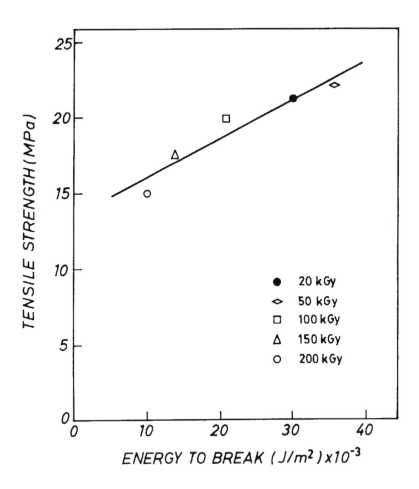

Figure 13. Plot of energy to break vs. tensile strength.

The dependence of tensile strength on gel content (%) of three sets of irradiated samples (i.e., EVA/TMPTMA, EVA/TAG, PE/TAC) is shown in Figure 14. In the case of EVA/TMPTMA and EVA/TAG systems the tensile strength attains an optimum value after an initial steep increase. The effect of gel content on tensile strength for PE/TAC system is, however, marginal. This can be explained from the molecular structures of the base polymers. Due to I effect (Inductive effect) of the ester group present in EVA, the α-hydrogen atoms (with respect to -OCOCH$_3$ group) are more labile and easily knocked out during irradiation even at lower dose, leading to the formation of large crosslinked structure, and hence increase in the gel content as well as tensile strength. At higher radiation dose the crosslinked structure breaks down to form small network structure. As a result the tensile strength decreases with gel content. In the case of PE, the tensile strength changes marginally with gel content due to lack of the above structural feature and both crosslinking and chain scission take place simultaneously over the whole range of radiation doses.

Electrical Properties

As the polymers are irradiated, a large number of free radicals are formed, which initiate crosslinking reactions. As a result the polymers form large network structure. The movement of individual chain segment then gets restricted and as a result the dipole orientation polarization becomes low. Since dielectric constant and loss factor depend very much on orientation polarization [20], these properties decrease on irradiation. Figure 15 shows the results for the EVA-TMPTMA system. However, in addition to this crosslinking, aerial oxidation takes place on irradiation, leading to the formation of polar carbonyl groups. The extent of this oxidation increases with the radiation dose. The increased concentration of carbonyl group at higher radiation dose leads to increased degree of dipole polarization. As a result the loss factor for EVA in Figure 12(a) increases beyond 100 kGy. The marginal change of ε' beyond 20 kGy is due to lack of chain movement because of crosslinking and grafting and balancing effect between various reactions. The effect of TMPTMA level on dielectric constant and loss factor at three different temperatures (namely 30°C, 100°C and 140°C) presented in Figure 15(b). It indicates that the loss factor decreases when TMPTMA (1%) is added to the system. But further increase of TMPTMA increases the tendency to crosslinking and grafting leading to a decrease in loss due to hindrance in chain mobility. However, as TMPTMA itself is highly polar, it contributes significantly to dielectric loss at higher dose. As a result this helps also in balancing the decreasing trend of loss factor. The same argument is true when the vinyl acetate content in EVA is varied [18]. In the case of polyethylene, which is a non polar plastic, incorporation of TMPTMA/TAC or increased irradiation dose increases the polarity. However, the gel fraction is also changed with these factors. The net effect is very similar to the results discussed above. The maxima or minima in the dielectric loss vs. temperature plot will depend on the structure of the polymer, radiation dose, type and level of multifunctional monomer.

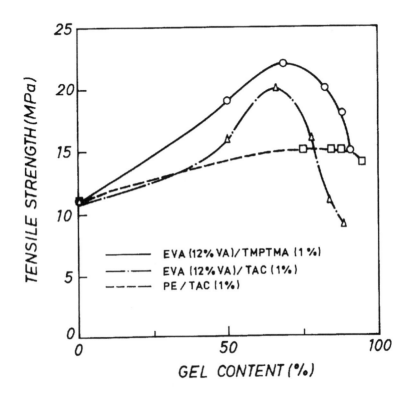

Figure 14. Variation of tensile strength with gel content of EVA/TMPTMA, EVA/TAC and PE/TAC.

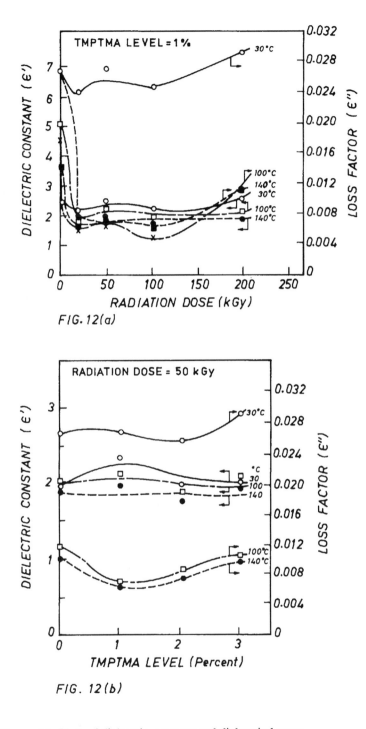

FIG. 12(a)

FIG. 12(b)

Figure 15. Plots of dielectric constant and dielectric loss vs.
(a) radiation dose and (b) TMPTMA level.

Dynamic Mechanical Properties

Viscoelastic properties are important in many practical applications of polymers. It is well known that the loss tangent (tan δ) is low and near zero when the material has elastic properties, and assumes higher values (δ near 90°) when the polymer has viscous properties. The temperature scan of tan δ reveals three transitions: γ (in the temperature range -140 to -100°C), β(-100 to -50°C) and α(-15 to +105°C) for PE and EVA copolymers (Figure 16). The height of tanδ in γ-relaxation due to movement of the main chain of the polymers increases in irradiated samples. The corresponding temperature of $(\tan \delta)_{max}$ shifts to higher value as the radiation dose is increased. The molecular defects resulting from irradiation impede the motion of the molecular chains and increase the potential barrier which causes the γ-mechanical relaxation process [21]. Introduction of TMPTMA into EVA causes a decrease in the $(\tan \delta)_{max}$ γ-relaxation region due to crosslinking. A similar behavior is observed for LDPE [21,22].

The β-relaxation comes around -100 °C to -50 °C and is very weak due to short vinyl acetate side group, while it is more intense in polyethylene due to short and long branching [21]. At higher temperatures a sharp peak (namely α-relaxation) is observed for all unirradiated and irradiated samples. With an increase in irradiation dose and multifunctional monomer level, the $(\tan \delta)_{max}$ shifts to the lower temperature. The value of $(\tan \delta)_{max}$ increases with an increase in irradiation dose. The α-relaxation is related to the melting peaks and their temperature is governed by the crystallinity and the most probable crystalline thickness [21]. The crosslinking and grafting takes place on irradiation which modify the molecular structure and hinder the growth of the crystal. The α-relaxation peak height increases with an increase in vinyl acetate content due to decrease in elastic characteristic. The corresponding temperature shifts to lower value which indicates decrease in crystallinity due to more crosslinking with increasing vinyl acetate content [18].

APPLICATIONS

Electron beam irradiation can vulcanize or accelerate the vulcanization of some elastomeric compounds without heating them. The curing and grafting reactions take place at temperatures at which there is no heat induced softening. As a result the irradiated objects retain their shape, dimensions and also any volatile matter in the composition. This technique is, therefore, useful for curing products such as unsupported sheeting and electrical insulation that would soften and sag at elevated curing temperatures and articles that would loose essential volatile components at elevated curing temperatures. Radiation vulcanization of natural rubber latex (RVNRL) is also a useful technique for producing dipped goods of which the various types of rubber gloves are largest items [23]. The electrical properties are also not affected here because of the non-polar compounding ingredients.

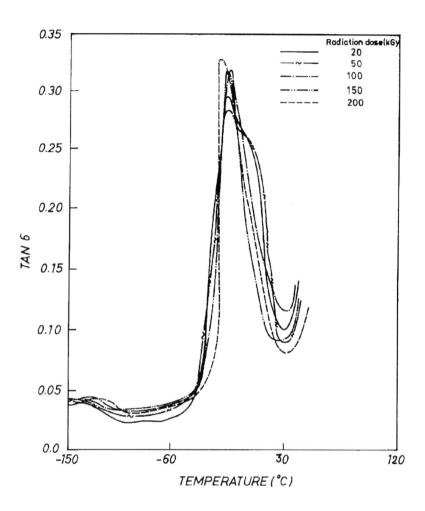

Figure 16. Temperature scan of tan δ of various irradiated samples.

Two applications are detailed below:

(a) Heat Shrinkable Cable Insulation

Jointing and termination are essential procedures for the efficient and proper transmission and distribution of power through cables. To extend the cable, frequent joints are needed. Also, the end terminations during storage need to be protected. Heat shrinkable and radiation crosslinkable polymeric materials may be used for this purpose.

Polymers are modified with the multifunctional monomers in an internal mixer and then extruded in the form of a tube. A typical recipe contains polyethylene with 4 parts of TMPTMA irradiated with 200 kGy radiation dose. These are then irradiated in the presence of electron beams followed by passing them through a tube expander. A typical expander contains five sections:

(a) Motorized pay-off section
(b) Cater pillar section
(c) Heating and expansion section
(d) Cooling section
(e) Pinch roll section

The heat shrinkable properties of these tubes are demonstrated in Table 4.

Table 4. Expansion and shrinking properties of polyethylene tubes.

Conditions	O.D (mm)	I.D (mm)	RT(mm)
After expansion	5.34	4.14	0.60
After shrinking	4.10	2.32	0.88

(B) Electron Beam Curing of Tire Components

By far the greatest potential for electron beam processing is in the manufacturing of tires for automobiles and trucks. The advantages are in the fast production rates and the ability to obtain improved dimensional stability of the components during the construction. Electron beam radiation may be used in the tire industry to procure or partially crosslink the components such as inner liners, body plies, side walls, chafer strips, inner strips and tread ply skims. Essentially, the real benefit lies in the utility of electron beam processing of tire components and the response obtained with the various elastomers to achieve greater form stability and retention of shape during assembly, molding and vulcanization. The properties of a typical electron beam irradiated halobutyl inner liner vulcanizates reported by Mohammed and Walker [5] are given in Table 5. Irradiation of the compounds improve green strength and fatigue properties, although there is a marginal decrease in tensile strength.

Table 5. Physical properties of halobutyl based inner liner vulcanizates.

Radiation dose (kGy)	0	10	20	30	50
Chlorobutyl					
Hardness (shore A)	48	44	44	43	40
Mod. at 100% elongation (MPa)	0,9	1,0	0,9	0,9	0,8
Mod. at 300% elongation (MPa)	4,1	4,3	4,0	3,9	3,5
Tensile strength (MPa)	8,7	7,3	6,7	6,1	5,4
Elongation at break (%)	600	530	510	480	470
Peel adhesion at RT (kN/m)	2,2	4,3	3,9	3,0	2,4
Peel adhesion at 100ºC	1,7	1,1	0,6	0,5	0,3
Tack (kPa)	290	240	280	190	110
Fatigue (kCy)	170	115	20	15	15
ML 1+4 (100ºC)	52	60	87	95	110
Green strength (kPa/100%)	-	3	45	86	---

ACKNOWLEDGMENT

The authors are grateful to the Department of Atomic Energy for helping in the electron beam experiments at various stages and Dr. A.B. Majali. BARC, Bombay and Dr. V. K. Tikku of Nicco corporation Ltd, Calcutta for continuous encouragement.

REFERENCES

1. H.F. Mark, N.M. Bikales, C.G. Menges, J.LKroschwitz. eds.. Encyclopedia of Polymer science and engineering, Vol.4 John Wiley & sons. New York,1988. p 418.
2. S.S. Ramamurthi, S.C. Bapna, H.C.Soni and K.Kotaiah, Proc. Indo-USSR seminar on Industrial applications of electron accelerator, Bombay, Bhabha atomic research center, Bombay Vol 2, p 58 (1988).
3. A.K. Bhowmick, Invited lecture, NAC-95, Bhabha Atomic Research Center. Bombay, India. 1995.
4. A. Charlesby, Plastics and Rubber processing and Application ,2,289 (1982).

5. S.A.H. Mohammed and J.Walker, Rubber Chem. Technol. 59, 482 (1986).
6. A. Charlesby. Atomic Radiation and polymers, Pergamon press, Elmshord, New York 1960.
7. W. Perkinson, G. Bopp, D. Binder and J. White. J. Phys. Chem., **69**, 828 (1963).
8. M. Tsuda and S. Oikawa, J. Polym. Sci., 17, 3759 (1979).
9. A.K. Bhowmick, M.M. Hall and H. Benarey, Rubber Products Manufacturing Technology. Marcel Dekker.N.Y. p.388 (1994).
10. K. Krishnamurthy, Proc. Indo-USSR Seminar on Industrial Applications of Electron accelerator, Bombay, Bhahbha Atomic Research Center, Bombay. Vol 2, p 1 (1988).
11. G. Socrates, "Infrared Characteristic Group Frequencies", Wiley Interscience., New York, 1980.
12. S.K. Datta, A.K. Bhowmick, R.S. Despande, A.B. Majali and T.K. Chaki, Polymer 37, xxx (1996).
13. S.K. Datta, A.K. Bhowmick and T.K. Chaki. J. Appl. Polym. Sci (in press).
14. S.K. Datta, A.K. Bhowmick and T.K. Chaki, Radiat. Phys. Chem, 45, xxx, (1995).
15. V.K. Tikku, G. Biswas, R.S. Despande A.B. Majali, T.K. Chaki and A.K. Bhowmick, Radiat. Phys. Chem 45, 829 (1995).
16. T.K. Chaki, S. Roy, R.S. Despande, A.B. Majali, V.K. Tikku and A.K. Bhowmick. J.
17. L.P. Nethsinghe and M. Gilbert, Polymer, **29**, 1935 (1988).
18. S.K. Datta, T.K. Chaki and A.K. Bhowmick, Rubber. Chem. Technol. **69**, xxx, (1996).
19. L.P. Nethsinghe and M. Gilbert. Polymer **30**, 35 (1989).
20. C.C. Ku and R. Liepins, "Electrical properties of polymers" Hanser Publishers, Munich-Vienna, New York, p. 21 (1987).
21. A.R. Greus and R.D. Calleja. J. Appl. Polym. Sci, **37**, 2549 (1989).
22. C.R. Aschcraft and R.H. Boyd. J.Polym. Sci, Polym. Phys. **14**, 2153 (1976).
23. S.M. Rao and K.M. Kulkarni "Isotopes and radiation Technology in Industry" (Proc. of the International Conference on Application of radioisotopes and Radiation in Industrial Development, ICARID-94, Feb 7-9, 1994, Bombay. India.

8

New Developments in Reactive Processing of Thermoplastic Composites

Vera Khunova and M.M.Sain
Department of Plastics and Rubber, Faculty of Chemical Technology,
Slovak Technical University, Slovak Republic
Pulp and Paper Research Center, University of Quebec,
Trois-Rivieres, PQ, Canada

INTRODUCTION

The effect of m-phenylene bismaleimide on the performance of calcium carbonate, magnesium hydroxide, talc and zeolite filled polypropylene was studied. Unlike conventional catalytic methods for composite modifications, the proposed mode of reactive processing is free of any initiator system. It was found that in the selection of processing conditions, temperature played a significant role in the preparation of such composites. A positive effect of the bismaleimide compound was obtained for a processing temperature above the decomposition temperature of bismaleimide. The effectiveness of modifier is more evident with higher filler content. A high stress material can be prepared even if the filler is the dominant phase in composites. The chemical composition of the modified composite and interaction between polymer and filler was correlated to its mechanical strength. The method used an innovative approach to improve interfacial interaction between the polymer matrix and filler by promoting a temperature sensitive chemical reaction via the reactive functionality of the modifier. The filler dispersion has been dramatically improved and the self-initiated chemical reaction between modifier, filler and polymer improved the mechanical strength of composites. Thus by using the *in-situ* method for composite modification, high stress materials can be prepared even if the filler is the dominant phase in the composite.

Significant development of particulate polymer composites which is drawing world-wide attention nowadays, is oriented to prepare easily processible, economically effective and ecologically viable materials with improved application properties.

While unique properties, such as dimensional stability or increased hardness and modulus, are the usual motivation for exploiting particulate filled composites, special attention must be paid to other mechanical properties, such as yield and ultimate strength and fracture toughness [1]. Due to the low adhesion between non-polar hydrophobic polymer matrix and hydrophilic filler surface, the interfacial debonding is very frequently the first step of failure in these materials [2]. Mainly for this reason the effective application of particulate composites is first of all determined by the interfacial interaction between polymer and filler.

Several methods have been reported to improve interfacial interaction in polymer-filler composites [1-7]. Most of those methods use polymeric interface modifiers and/or multifunctional reactive monomers in conjunction with a suitable catalyst system to initiate the reactive modification process. Among those, multifunctional reactive monomers are finding more and more applications in recent years to modify the properties of particulate as well as fiber-filled thermoplastic composites [5-8]. Both maleic anhydride [7,8] and bismaleimide derivatives [9] had been used in the past to improve the strength and thermal resistance of composites and polymer blends. The use of these reactive additives are mostly combined with a controlled catalytic process initiated by radical, plasma or UV. As discussed in the past [10] these initiation processes have several disadvantages that influence the ultimate performance of the composites. Firstly, the catalytic reaction process is never complete and the reaction conversion is always less than 100%. This incomplete conversion leaves behind variable amount of residual monomer and catalyst in the composite matrix which are detrimental to composite life and properties. For example, a residual free radical initiator in the composite might deteriorate the aging resistance of the composites. On the other hand, a residual peroxide initiator might act as a crosslinking agent for certain other polymers [11]. Thus, technologies are developed to minimize the residual initiator concentrations in composites. In earlier investigations they have advantageously used the method to modify powdered PP in solid phase by promoting surface reactivity at a relatively low temperature and then used this highly effective modifier to improve the properties of polyolefin-based inorganic composites [12,13].

More recently, reactive extrusion under an inert atmosphere was found to be very effective to minimize the side effects of composites prepared by in-situ reactive processing [14]. The economic feasibility of such processes are recently questioned due to the involvement of multistage separation processes of residual catalyst or for providing an inert atmosphere to minimize the side effects.

It is known that many multifunctional monomers are prone to thermal initiation. However, this method is very complex for composite preparation because a combination of high temperature and high shear frequently leads to thermo-mechanical degradation of the polymer phase resulting in deterioration of mechanical and thermal properties and it is particularly so if the filler loading is very high. Thus a proper selection of polymer, filler and reactive monomer may not be enough to carry out a chemical reaction unless a suitable process condition to minimize those side reactions is found.

Reactive processing is now being recognized as an efficient approach to performing a wide range of chemical reactions and modifications of polymers. Many of these processes are carried out in an extruder and require the use of organic peroxides, e.g. the production of controlled rheology polypropylene and the grafting of vinyl monomers onto polymers. The graft modification process involves the use of organic peroxides to create a reaction side on the polyolefin backbone. The organic peroxide type and the processing conditions play an important role in the grafting efficiency and the rheologioal properties of the final product [15] .

The direct addition of a reactive monomer, namely, maleic anhydride, in the compounding machine apparently does not have much effect on polymer-filler interaction [16]. Addition of maleic anhydride in PE directly in the extruder along with peroxide activator and/or dibutylphthalate and $CaCO_3$ improved mechanical properties of this polymer composite and the improvement is very much dependent on the processing time. Some improvement in the case of above composite is obvious because PE can undergo crosslinking reaction with peroxide catalyst. The same reaction is apparently impossible for polypropylene based composites. It is because PP has tertiary carbon atom which readily undergoes chain scission reaction with peroxide catalyst. Other disadvantage of direct malleation of polyolefin composite is residual maleate and peroxide concentration. Free residual maleic anhydride (MAH) in composite formulation generates toxic fumes during processing . Moreover, any unconsumed catalyst in the maleated polyolefin composite negatively influences the mechanical strength and causes deterioration of the thermo-oxidative properties of composites as well. Therefore, the direct addition of maleic anhydride in the composite processing reactor is not advisable. It is particularly so if the polymer is very much prone to peroxide degradation reaction. Development in this line gave rise to a unique product , maleated polyolefin. Maleated polyolefin is now widely used as a reactive polymer in polyolefin composites.

Because the preparation of maleated polyolefin involves a separate stage different from a composite processing stage the process does not enjoy the maximum economic benefit that could be obtained by using a reactive monomer directly in the composite processing stage.

The effect of heat and catalyst on a multifunctional bismaleimide derivative has been discussed [17]. It was observed that the m-phenylene bismaleimide (BMI) undergoes catalytic decomposition even below 150°C, but a similar decomposition process was also evident at 190-200°C in absence of a catalyst. Therefore, it is possible that BMI can undergo chemical reaction even in absence of a radical initiator. Keeping this fact in mind, the effect of the m-phenylene bismaleimide on the reactive processing of inorganic-filled polypropylene was studied over a temperature range where BMI easily undergoes decomposition reaction with the generation of reactive radicals in absence of a peroxide initiator.

In this context it is worth-mentioning the role of reactive bismaleimide in composites based on polypropylene and inorganic or organic fillers. Presumably, the reactive bismaleimide improves the quality of interfacial zone in polypropylene based composites by three different mechanisms: by improving the wetting of the filler surface with the matrix component, by establishing mechanical as well as physico-chemical interactions between filler and matrix and finally, by solidification of a matrix parallel to the formation of an interfacial linkage [18]. It is expected that the concentration of reactive monomer, the processing temperature and the processing time will influence the quality of the interface formed and, hence the strength of the composite.

This chapter discusses the effect of addition of a reactive monomer, namely, m-phenylene bismaleimide (BMI) in the composite processing stage on the processing and end-use properties. Various particulate fillers were used for making composites and their properties were compared with respect to processing and end-use properties.

POLYOLEFINS AND FILLERS USED IN COMPOSITES

Before examining the properties of polyolefin-filler composites, it is appropriate to consider the characteristics of the polyolefin and filler components. This is, of course, because the properties of a composite depend on the characteristics of its component parts.

Among polyolefins, polypropylene has recently become an attractive candidate for many engineering applications. New polymerization processes introduced in the last decade make polypropylene one of the most favorable matrices for high-volume composites and blends. Relatively low priced, excellent chemical resistance, good processibility and the possibility to modify, it is spreading into automotive, land transport, home appliances and other industries. The mass application of PP in automobiles, mass transport vehicles, household appliances and the construction industry requires stiffness and fracture toughness similar to those of engineering plastics (ABS,

PC, etc.) [19]. Polypropylene is of much interest for composite manufacturing because of its high tensile strength, impact strength, flextual modulus and softening point. The mold shrinkage of polypropylene is less than that of the high density polyethylene. Table 1 furnishes some mechanical properties of polypropylene homopolymer.

Table 1. Selective properties of polyolefins.

Property	PP homopolymer	HDPE
Tensile strength, MPa	36	26-3 27-34
Elongation at break, %	350	500
Flextual modulus, MPa	1310	-
Brittle temperature, C	+15	-
Softening point, C	145-150	-
Hardness, (R-scale)	95	-
Impact strength, kJ.m-2	24.5	12.9
MFI, g/10 min	8.7(2.16 kg/230C)	1.5(2.16kg/190C)

The action of particulate fillers on a thermoplastic is dependent on factors that can be classified as extensity, intensity and geometrical factors. The extensity factor is the amount of filler surface area per m^3 of the composite in contact with the plastic. The intensity factor is the specific activity of this solid surface per m^2 of interface, determined by the chemical and physical nature of the filler surface in relation to the plastic. Geometrical factors are the structure including shape of filler (anisotropic such as such as lamellar, plate, needle and isotropic such as spherical), their particle size and size distribution as well as porosity. Among those the chemical nature of the filler surface plays most vital role in determining the degree of plastic-filler interaction.

Particulate fillers widely differ in filler characteristics. Depending on their physical and chemical nature the homogeneity of composite is very much influenced by filler. For example, particulate calcium carbonate differs chemically from talc; again talc and $CaCO_3$ are geometrically different from that of zeolite . Likewise, the chemical composition of hydrated magnesium hydroxide is different from talc., $CaCO_3$ or zeolite. Therefore, it is expected that those particulate fillers will act differently for a given thermoplastic. For example, fillers introduced into the polymer matrix in the form of micro-ground powder are usually difficult to disperse. Thus, the end use properties of composites are directly connected to the quality of the filler dispersion and physico- chemical interaction between filler and polymer matrix.

While unique properties, such as dimensional stability or increased modulus, are the usual motivation for exploiting particulate filled composites, special attention must be paid to other mechanical properties, such as yield and ultimate strength and fracture toughness, since they are often degraded by the presence of fillers [20].

In this experimental investigation multiple composite systems containing polypropylene and four different particulate fillers were selected for evaluation. Composites were prepared with and without additives to compare the effect of different chemical modifiers.

PREPARATION OF COMPOSITES

Thermoplastic composites, in general, have been prepared commercially by melt mixing and molding. Melt mixing avoids problems of contamination, solvent or water removal, etc. In general, internal mixer and twin screw extruders are suitable for making thermoplastic composites. However, for the purpose of this chapter, emphasis will be given only on the laboratory scale internal mixer, i.e., Brabender mixer, because the variability of this processing equipment very well simulates the industrial conditions for reactive processing of composites.

Unmodified Composites Prepared by Simple Melt Mixing

Unmodified composites of plastic and fillers are prepared by mixing plastics (unstabilized and stabilized) and filler for a given time period at the melt temperature at a roller speed of about 40 rpm. However, there are occasions when changes in rotor speed are of value. The appropriate mixing temperature and time depends on the nature and volume of the filler. The temperature must be high enough that the plastic and filler mix is molten and flows easily enough for mixing. At the same time, temperatures should be as low as possible to avoid rapid oxidative degradation. We have found that an initial experimentation is required to find a time and a temperature range for optimization of processing and end-use properties. Thus initial experimentation has been done to optimize the time and temperature for compounding. The temperature was varied from 180 °C to 220 °C and compounding time was between 3 to 8 minutes.

The above mixing cycle can be scaled to a Warner-Pfleider mixer or twin screw extruder. After being chopped, the extruded pellets or mixed crumb can be extruded again or pelletized, and injection molded, compression molded, etc.

Reactive Mixing of Composites

The reactive processing has been applied to many plastic-filler combinations in our laboratory. It can be described as follows: Plastic and filler are first melt-mixed in the same way as that described for unmodified composites. After sufficient melt mixing to form a well mixed composition, reactive monomer, i.e., m-phenylene bismaleimide was added. It is important to note that no initiator was added to the system. Reaction then occurs while mixing continues. It is convenient to follow the progress of reaction by monitoring mixing torque or mixing energy requirements during processing. After the mixing torque goes through a maximum it is advisable to continue the processing for some additional time to improve the flow property of the composite. The reactive mixing was carried for various temperature and time to optimize the reactivity of monomer in relation to polymer matrix-filler interaction. After discharge from the mixer the compositions are handled in the same way as for the unmodified composites. Typical mixing and molding temperatures are given in Table 2.

Table 2. Mixing and compressing molding parameters.

Composites	Mixing Temp., ^{o}C	Molding Temp., ^{o}C
PP-Filler (CaCO$_3$, talc, zeolite, Mg(OH)2	170-180	200
PP-Filler (iridization),	200-210	220-230

Optimization of Composition and Processing Parameters

Statistical experimental design and response analysis are very useful tools for optimizing product performance. In our investigation we have used this design to minimize the number of experiments as well as to obtain custom designed composites with required mechanical, chemical and/or thermal properties. In optimizing composite properties, reactivity and process parameters designed experiments were used. A major process parameter which needs frequent optimization is processing temperature and the other optimization parameters are usually polymer to filler ratio and amount of reactive additive. Two optimization experiments were necessary to justify the influence of compounding temperatures on the strength of bismaleimide modified composites. Both the designs used were two factorial rotatable central composite ones with five levels of coded variables. The concentration of BMI modifier was varied from 0 to 3 wt% and the filler concentration was varied from 0 to 60 wt%. The repeatability of the results was tested by analysis of variance (ANOVA) and the adequacy of the regression equations for each dependent variable was tested by unbiased F-test results.

For comparison the results of maleated PP modified composites were also prepared in the same way as that utilized for BMI modified composites.

The mechanical properties of the composite sheets made in laboratory are tested according to ASTM methods. The impact strength, however, was measured at -20^0C using Charpy impact tester. All these test results simulate the properties of end product. Besides physical and mechanical properties the chemical characteristics of modified and unmodified composites can be very well correlated from Infrared spectroscopy and ESCA analyses data.

PROCESSING   BEHAVIOR

In general thermoplastics are shear thinning. An increase in mechanical shear results in a decrease in the melt viscosity and a reduction in torque. The increase in the process temperature above the melting point of thermoplastics has a similar effect on their stress-strain behavior. Generally, a reduction in torque with increasing shear or temperature during processing is attributed to a decrease in viscosity resulting from molecular orientation and/or breaking down of the polymer chain. Therefore, torque can be used to optimize the process condition. In unmodified composites it is generally difficult to define a critical torque value because the torque progressively decreases with increase in processing time and temperature after a short rise in torque during initial mixing period. Unlike unmodified composites, reactive processing of composite shows a definite trend in the torque -temperature, torque-shear rate or torque-time behavior. Figure 1 shows the effect of temperature on mixing torque of a mixture of $Mg(OH)_2$, and PP. It is evident that the mixing torque increases with increasing temperature up to 250 °C for a mixing time of about 3 minutes. On the other hand, the mixing torque decreases progressively with increasing processing temperature if the processing time is increased to about 5 minutes. An increase in torque with increasing temperature for a given mixing time and shear rate essentially indicates a chemical interaction during processing leading to a shear thickening composite. The optimum reaction time and temperature for a given shear rate during composite processing therefore can be obtained from the peak of the torque versus temperature plot.

Similar to processing condition the composite composition including the optimum concentration of reactive monomer or polymer can be estimated from torque or viscosity behavior of the composite during processing. Figure 2 gives another example of the effect of a reactive additive concentration on the viscosity of the melt. The increase in the additive concentration increases the viscosity of the melt.

In general, for a unmodified composite the viscosity decreases with increase in the polymer concentration. During reactive processing of composites the viscosity behavior of composite changes with the addition of a reactive component even in trace quantity. Thus, the addition of MAH-PP

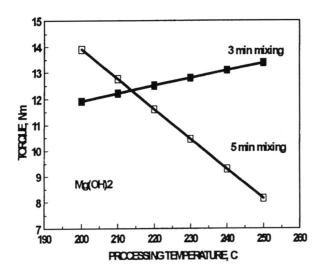

Figure 1. Effect of processing temperature on mixing torque.

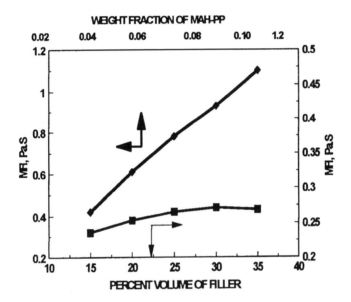

Figure 2. Effect of composition on flow property.

in a mixture of PP and $CaCO_3$ increases the viscosity of the melt. This rise in the viscosity can be correlated to a change in chemical composition of the composite during processing. Thus a reactive process in composite preparation can be monitored from the torque-viscosity data of the melt during processing. A similar trend in the viscosity vs filler loading is expected for BMI modified composites during processing at high temperature, i.e., 210 °C.

PROPERTIES OF THERMOPLASTIC COMPOSITES

Properties of Unmodified Composites

The properties of heterogeneous composites are determined by a number of factors. The main factors are (1) the material properties of the polymer and filler, (2) the proportions of polymer and filler, (3) the phase morphology and (4) the nature of the interface. Table 3 gives the tensile strength of the composite containing 40 wt% $CaCO_3$ for two different temperatures and three different times of compounding.

Table 3. Effect of processing condition on tensile strength of composite*.

TENSILE STRENGTH, MPa				
Variable Mixing Temperature, ^{0}C		Variable Mixing time, min		
180	210	4	6	8
26.6+0.5	23.6+2.0	26.4	26.7	26.5
Mixing time 4 min		Mixing temperature 180^{0}C		

* from Khunova and Sain, 1995(a); Huthig & Wepf Verlag, Zug [21].

Results of Table 3 shows that the processing time has negligible effect on the performance of the composite within the chosen time range and for a given processing temperature of 180 °C. Thus the optimum processing condition for $CaCO_3$-filled PP composite was found to be 180 °C and 4 minutes. Figure 3 illustrates the tensile property of heterogeneous composites made from a combination of thermoplastic with three different fillers. The tensile strength monotonously decreases with increase in filler content and depending on the nature of the filler the tensile values are different.

Reactive Composites with Maleimide

In contrast to unmodified composites, the BMI modified composite showed a dramatic change in the mechanical strength with changing processing temperature. This is evident from Figure 4. The strength can be improved by about 30 to 40% over unmodified composite if BMI is introduced in the composite even at a low processing temperature, i.e.,

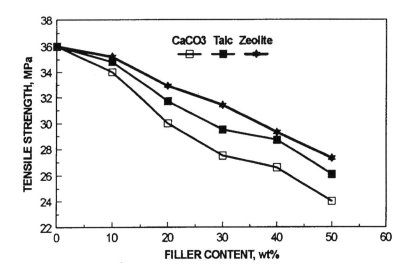

Figure 3. Effect of filler loading on tensile strength of PP composite.

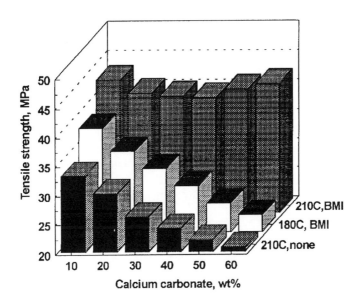

Figure 4. Effect of filler loading and processing temperature.

180 °C. Moreover, the relative improvement of tensile strength over unmodified composite increases with increasing filler loading. An increase in process temperature from 180 °C to 210 °C in the presence of BMI resulted in a surprising improvement of the strength performance thermoplastic composites. About two fold increase in the tensile strength has been achieved even if the filler is the dominant phase. It is then clear that new process to prepare composite by reactive mixing without initiator system would produce composites which are free from environmental hazards, such as decomposed products from conventional initiators, peroxides, azo-nitriles, etc. Moreover those composites are also free from residual initiator systems and therefore, they are expected to be more resistance to thermal and auto-oxidative degradation. A very similar result was also obtained with zeolite-based PP composites [21].

Figures 5 and 6 are the response surface diagrams showing the effect of filler content and BMI concentrations on tensile and impact properties of composites [22]. All composites were processed at 210 °C. The increase in BMI content progressively increased the tensile strength of the composite and the positive influence of BMI is more evident with high $CaCO_3$ concentration in the composite. This fact confirms that the decomposition of BMI plays a significant role to improve the interfacial interaction either by improving the surface wetting of filler or by interacting physically or chemically at the filler-matrix interface. The role of BMI as an interface modifier has been well explained in our earlier study where wood fiber was used as a filler component [17]. It is thought that a similar interaction between PP and BMI occurred in the present case.

In BMI modified composites, the stress transfer in the polymer matrix is improved even below the glass transition temperature. This suggests that the BMI helped to distribute stress from the polymer matrix to the filler particles by forming an interfacial bridge between them and improving impact properties. Similar mechanism was proposed for a wood -fiber-filled PP composite [18]. The measurement of glass transition temperature by time domain reflectometry (23) indicated a marginal increase in the glass transition temperature for PP in BMI modified PP. This fact also suggests a possibility of formation of interfacial bridge.

Table 4. Binding Energy of C, N, and O Core Levels Obtained from ESCA.

Samples	PP	BMI Treated PP[1]
Atomic ratio		
O/C	0.012	0.030
N/C	0	0.011

[1] PP was treated with BMI at 210⁰C for 4 min without addition of filler; from Khunova(1995a) [21].

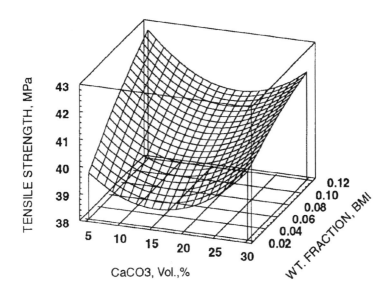

Figure 5. Effect of filler and modifier concentration on tensile strength of PP composites; from Khunova (1995)b [22].

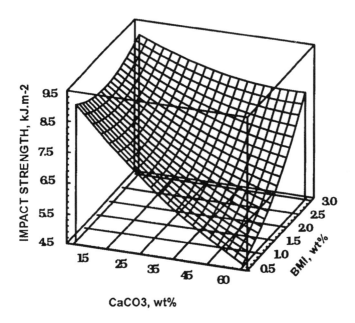

Figure 6. Effect of filler and modifier concentration on impact strength of PP composites; from Khunova (1995)b [22].

The ESCA results of PP and 3 wt% BMI treated PP at 210 °C are summarized in Table 4. The atomic ratio of O/C and N/C obtained from ESCA study indicated an increase in O/C ratio in BMI treated PP. Again, a nitrogen peak near 400 eV of the ESCA range for BMI modified PP was evident from Figure 7. Thus, the results of ESCA analysis suggest a surface bonding of BMI with PP. Hence the improvement of the tensile strength and impact strength of BMI modified composite is associated with some change in the interfacial tension between PP and $CaCO_3$.

Similar to $CaCO_3$, the incorporation of magnesium hydroxide in the PP matrix also improved the mechanical strength of the composite at higher processing temperature.

Results demonstrated in Figure 8 confirms that the optimum processing temperature for obtaining highest tensile strength for a given composition of the composite would be in the range 210 to 225 °C. The impact strength of composite does not show any improvement with increasing process temperature. It is then advisable to use a variable process temperature to obtain a custom-designed composite with predetermined tensile and impact properties.

Beside process temperature, the compositional variation of $Mg(OH)_2$-filled composite demonstrates a significant difference in properties. The reactive monomer, BMI, concentration is very critical to obtain a composite with optimum strength and impact resistance. From Figure 9 it is evident that the concentration of reactive monomer would be around 6 wt% for a composite with good impact strength and reasonable tensile strength. However, the tensile strength can be improved further at the expense of impact property.

If we compare the properties of composites obtained from different fillers we could see a significant drift in their properties. We have used four different fillers and compared their processing behavior and end-use properties as a function of processing conditions and product composition. Figures 10 compares the tensile properties of composites filled with different fillers.

Results demonstrate the dependence of product composition on tensile strength. It indicates an optimum composition for tensile strength in talc-filled composite. An interesting trend has been noticed if one compares the effect of BMI on the tensile strength for two extreme filler loading conditions. The strength of PP marginally increased with increasing BMI concentration for practically unfilled composite. This fact predicts a different mechanism by which BMI might function in composite. Presumably, PP might be crosslinked by BMI in a very lightly filled composite and such chemical crosslink, if any, might help to improve the strength of PP as shown in Figure 10. If the chemical crosslinks really formed, it should be indicated by a residual insoluble PP content in a selective solvent extraction

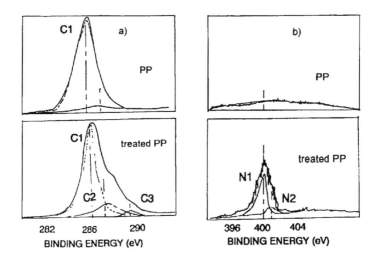

Figure 7. ESCA spectra of untreated and BMI-treated PP; (a) deconvoluted
C_{1s} spectra, (b) deconvoluted N_{1s} spectra;
from Khunova (1995a) [22].

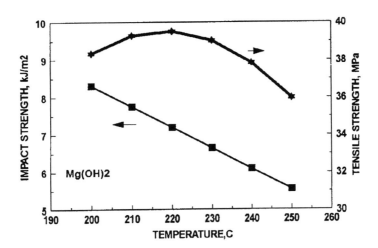

Figure 8. Effect of temperature on mechanical properties of Mg(OH)$_2$-filled
PP composites.

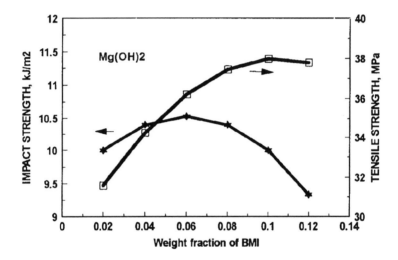

Figure 9. Effect of modifier concentration on Mg(OH)$_2$-filled composites.

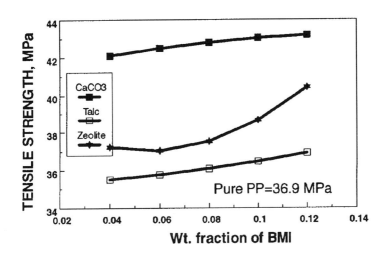

Figure 10. Effect of filler content on tensile strength of 3% BMI modified composite.

of the same composite sample. Laboratory test results based on a hot xylene extraction for 18 hours, however, did not leave any solid residue. Thus solvent extraction test result failed to support the reason for an increase in the strength of very lightly filled PP composite. Nevertheless, it can be argued that some minor chemical interaction did occur during the mixing and molding processes which were not strong enough to withstand the extreme thermal condition imposed during solvent extraction process.

A difference in the absolute value of the ultimate properties was observed if the type of filler component was changed. The tensile strength of zeolite-filled composite demonstrates a maximum for the highest concentrations of BMI and zeolite in the chosen experimental range. This fact indicates that degree of interaction between PP and zeolite is directly proportional to the concentration of filler for a BMI concentration of about 2.5 wt%.

Table 5. Impact strength of filled composites.

Filler type	wt % filler	Reactive modifier	Impact strength, kJ.m-2
$CaCO_3$	40	none	4.2
	40	BMI	9.7
Zeolite	40	none	4.9
	40	BMI	7.2
$Mg(OH)_2$	60	none	2.4
	60	BMI	9.3
Talc	40	none	8.1
	40	BMI	9.1

The impact strength of the composite has a negative influence on the filler concentration. In general, the more the loading, the less the resistance to impact at low temperature. The absolute value of the impact strength, however, has a marginal positive influence on the BMI concentration compare to that of the unmodified ones [21,22]. Table 5 furnishes the impact strength values of several composites at low temperature (-20⁰C). The improved impact strength of reactivitely modified composites can be attributed to better stress distribution either due to the changes in the phase crystallinity and/or due to the formation of chemical linkages.

The impact strength of a thermoplastic matrix is dependent on the molecular characteristics including polymer chain length and degree of branching. Usually, a minor interaction between chain molecules might help to facilitate the energy transfer due to the application of sudden impact stress, a strong chemical and/or physico-chemical interaction between PP chains could prevent the uniform stress distribution throughout the composite matrix due to the loss of the chain mobility. Therefore, experiments were carried out to investigate the nature of interaction in BMI-modified talc and zeolite composites. Figure 11 compares the IR spectra for BMI modified composite with that of unmodified ones for two different BMI concentrations. All samples were cold extracted in tetrahydrofuran (THF) to extract unreacted BMI from composites. The difference spectrum apparently shows the development of a peak near 1718 cm^{-1}. The peak can be separately attributed to a cyclic imide peak arising from BI, a carbonyl group due to oxidation of PP, or an ester bond between BI and fiber, or it could be a cumulative effect of two or more of these factors. Again a model experiment using a peroxide catalyst during modification of composite with BI indicated an additional oxidation peak at 1736 cm^{-1}. This observation suggests that the peak at 1718 cm^{-1} may be due to imide linkage, not due to oxidation of PP. The spectrum of modified composite also shows a small absorption band at 841 cm^{-1} (not shown in figure) attributed to vibrational band of crystalline PP. Apparently, both PP and the imide moiety of BI are chemically bonded [5]. Moreover, a peak at 1500 cm^{-1} is likely to be responsible for the stretching vibration of aromatic ring in BI. It is also found that the peak intensity increases with increase in BI concentration. This observation also supports our earlier presumption that a chemical reaction between PP and BI as proposed by Said and Cacti [7].

USES OF REACTIVELY PROCESSED COMPOSITES

A major application of thermoplastic composites of this type has been in exterior automotive body parts such as filler panels and bumper covers. These composites are also used for interior parts of automotive body such as car interior side panels, front panel, door interior, etc. For constructional applications these composites can suitably replace interior housing panels including patio and window panels. They can also be used for injection molded furniture.

For electrical and electronic industries these composites can be used for instrument panels, housings etc. There are several other applications are cited in literature [25,26].

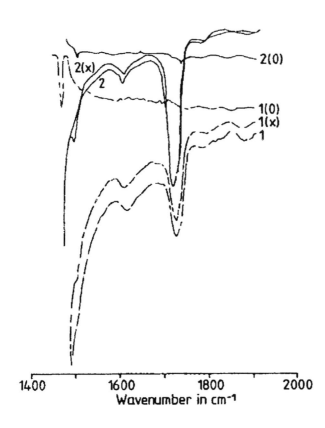

Figure 11. IR spectra of unmodified and BMI-modified PP; 1 and 1(x) are
the spectra of unextracted and extracted composites modified with
0.7 wt.% BMI, respectively, and 1 (0) is the difference spectrum
of 1 and 1 (x); 2 and 2(x) are the spectra of unextracted and
extracted composites modified with 3 wt.% BMI, respectively,
and 2(0) is the difference spectrum of 2 and 2(x); from Khunova
(1995)b [22].

REFERENCES

1. J. Jointure , A. Dianseimo , A.T.DiBenedetto; Polymer Engineering and Science, 32, (8) pp. 1394-1399, 1992.
2. B. Pukanszky, F. Tudos, J. Jancar, J. Kolarik; J. Mat. Sci. letters, 8, p. 1040-1042, 1989.
3. V. Khunova, V. Smatko, 1. Hudec and J. Beniska; Progr. Colloid Polym. Sci., 78, 188   (1988).
4. V. Khunova, M. M. Sain and 1. Sirnek; Polym.-Plast. Technol. Eng.,32(4), 299 (1993).
5. M. M. Sain , C. Imbert and B. V. Kokta; Die Angewandte Makromolekulare Chemie,   210 (3586) 33 (1993).
6. V. Khunova, M. M. Sain and Z. Brunovska; Polym.-Plast. Technol. Eng., 32 (4) 311   (1993).
7. M. M. Sain and B. V. Kokta; J. Adv. Polym. Technol. , 12 (2) 167 (1993).
8. J. Muras and Z. Zamorsky, Plasty a Kaucuk, 28 (5) 137 (1991) (in Czech).
9. I.Hudec, M. M. Sain and V. Sunova; J. Appl. Polym. Sci., 49 (3) 425 (1993).
10. R. J. Raj, B. V. Kokta and C. Daneault; Intern. J. Polymeric Mater., 12, 239 (1989).
11. R. G. Raj and B. V. Kokta; Die Angewandte Makromolekulare Chemie, 189 (3192)   169(1991).
12. V. Khunova and J.Zamorsky; Polym.-Plast. Technol. Eng., 32(4) 289-298 (1993).
13. V. Khunova, M. M. Sain and 1. Sirnek; Polym.-Plast. Technol. Eng., 32(4) 299-309   (1993).
14.10. D.J. Olsen, Proc. 50th Annual Tech. Conf., ANTEC 1991, Montreal, p. 1886.
15. P.A. Callais and R.T.Kazmierczak ; The Maleic Anhydride Grafting of Polypropylene with Organic perixides ANTEC 90, pp. 1921-1923.
16. M. Brauer, KJanichen, V. Mulller, G. Zeppenfeld; Plaste und Kautschuk, 35, 2, pp. 42-46,1988.
17. M. M. Sain and B. V. Kokta; J. Adhesion Sci. Technol., 7 (7) 743 (1993).
18. M. M. Sain, B. V. Kokta and D. Maldas; J. Adhesion Sci. Technol.; 7 (1) 49 (1993).
19. J. Jancar, A. DiBenedetto;The mechanical response of ternary composites of   polypropylene with inorganic fillers and elastomers inclusions, ANTEC 1993, pp. 1698-1700.
20. L.Anderson, R.Farris ; Polym. Eng. Sci. 28, 522, 1988.
21. V. Khunova and M. M. Sain; Die Angewandte Makromolecular Chemie 224 (3853)   p.9-20(1995)a.
22. V. Khunova and M. M. Sain; Die Angewandte Makromolecular Chemie 225 (3854) p.11-20(1995)b.
23. M. M. Sain and W. Otowski, unpublished work.
24. M. M. Sain and B.V. Kokta; J. Adhesion Sci. Technol.; 7(1) 49-61

(1993).
25.   D.M.Bigg, D.F.Hiscock, J.R.Preston, J.E.Bradbury; Polymer Composite, 9, 992 1988.
26. J.Jancar, A.T. Di benedetto; Journal Mater Sci., 29, 4651-4658, 1994.

9

Application of Polymer Technology to Metal Injection Molding (MIM) Processing

M. L. Foong and K. C. Tam
School of Mechanical and Production Engineering,
Nanyang Technological University,
Nanyang Avenue, Singapore 639-798.

INTRODUCTION

Metal injection molding (MIM) is a form of powder metallurgy process that is ideally suited for the production of small complex parts. Its main advantage lies in its ability to produce finished parts to net-shape, requiring little or no expensive secondary operations. Hence, it is extremely attractive for hard metals and high precision or high performance parts.

Although the MIM process has great potential and numerous advantages, much of the process technology has been kept in commercial secrecy, licensees and guarded by numerous patents. Despite the increase in scientific research into this area in the recent years, in particular by institutions like Rensselaer Polytechnic Institute, Penn State and Brunei University, there still remains much to be investigated in detail. One of the major technical challenges faced by the researchers lies in the formulation of a right binder system for optimizing the molding and debinding characteristics of the MIM process.

This chapter focuses on the application of polymers in the formulation of binder blends for MIM. A brief summary of the different classifications of binder blends is presented. The rheological, thermal and mechanical properties of a feedstock with EVA/beeswax binder is discussed to highlight the influence of binder-powder interactions on these properties.

METAL INJECTION MOLDING

The metal injection molding process consists essentially of the following steps:

i.   Preparation of powders;
ii.   Design and preparation of binder system;
iii.   Mixing to form a homogeneous feedstock;
iv.   Pelletising or granulation of feedstock;
v.   Injection molding in a closed die;
vi.   Binder removal (debinding);
vii.   Sintering.

The MIM process, illustrated in Figure 1, involves incorporating a binder into fine metallic powders. The binder serves a dual role of a transport medium allowing flow and packing of the powder in the mold cavity, as well as a medium to retain the shape of the molded part. The binder and powder is mixed in a mixer until a homogeneous mixture is obtained. The mixture is then pelletised to form a feedstock and subsequently fed into an injection molding machine. Next, the binder in the molded parts is removed in the debinding step. Debinding is achieved either by solvent extraction or thermal decomposition. Once the binder is entirely removed, the molded parts are sintered in a controlled atmosphere furnace or vacuum furnace to full density.

In view of the present trend of miniaturization, commercial and domestic products are progressively designed to be smaller, lighter and more compact. As a result, there is a growing need for smaller and more complex shaped parts. Moreover, factory automation requires products to be designed with ease of assembly. This inevitably requires the number of parts per product to be reduced and hence, the deliberate redesign of parts with multiple functions. Individual components become more complex in shape. Due to these various reasons, it is evident that a cost effective method to produce small and complex precision metal components in high run volumes for custom parts markets is very much in demand.

MIM, which possess such properties, offers a very attractive alternative over the conventional methods, such as investment casting, discrete machining and other conventional powder metallurgy process where secondary machining would be needed to create additional geometry.

Figure 2 shows the basic attributes of MIM, a combination of low cost, high performance and shape complexity. MIM would achieve low cost through ease of processing, low cost of capital equipment and elimination of secondary operations. It is capable of producing high performance parts through high final density and even the possibility of composites (e.g., metal matrix composites). Finally, it is capable of producing net shape parts of complex geometry with excellent surface finish. In addition, the process can

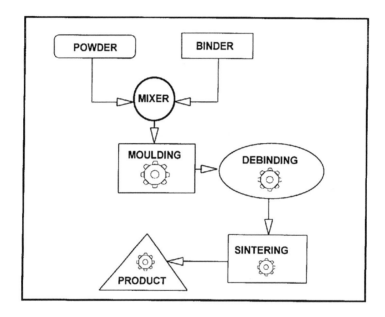

Figure 1. The MIM process route.

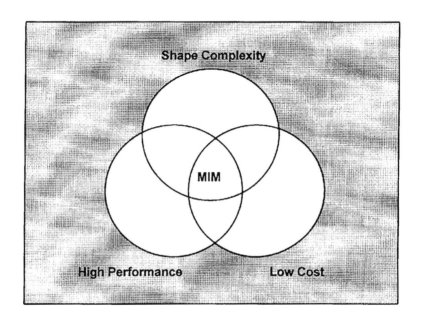

Figure 2. Optimal application of MIM [1].

be automated and is open to modeling via modification of modeling methods already established in thermoplastics molding [2]. As MIM's potential is being recognized, it has found applications in a variety of areas, including firearms, orthodontic devices, business machines/printers, computer disk drives, dental and medical instruments, household appliances, watches, jewelry and electronic packages. Billiet [3] reported that the number of organizations engaged in MIM research or production had risen from less than half a dozen in 1980 to well over 100 worldwide in the 1990s.

Despite its optimistic growth, the MIM/PIM industry growth is being hindered by various factors. Messier [4] classified these hindrances into non-technical or technical hindrances. The main non-technical hindrances include:

i.   *Lack of complete understanding of the process.* There is a lack of complete understanding of the MIM process. Those already involved in production tend to rely on a trial and error approach rather than a systematic study of the process. Much of the knowledge of the subject is derived from experience rather than understanding.

ii.   *Overprotection.* There is overprotection of the technology with patents, licenses and proprietary rights. Organizations involved in research in the process are generally unwilling to reveal or publish their findings .

The main technical hindrances would include the following points.

i.   *Empirical binder formulations.* Binder formulation is empirical and normally would work for a particular system but not others. There should be more in-depth understanding of the binder formulations based on scientific grounds.

ii.   *Limit on product size.* There is a limitation on the size of the parts which includes the wall thickness and overall weight. Oversize parts are vulnerable to defects and distortions. Moreover, the excessively long debinding time for large parts would render it uneconomical to produce by MIM.

iii. *Understanding of process.* Poor understanding of the actual process and shortage of technical personnel limits the powder selection, dimensional control, process optimization, process control and diversification of applications and materials.

For the success of MIM industry, further efforts would be required to promote the credibility of the process which would in turn promote the acceptance of the process. In view of this, thorough and systematic research to understand the underlying principles would be necessary. Only then would MIM be widely accepted and used in the manufacturing industry.

BINDER FORMULATION

The Role of the Binder

As previously mentioned, the chief role of the binder is essentially to carry the powder and allow flow and packing of the powder in the die cavity. Subsequently, the binder is also necessary for the green body to retain its shape before sintering.

The binder is the key component in MIM. It controls the mixing, molding and debinding operations. It also strongly influences the maximum powder loading in the feedstock, the green strength of the molded parts, the shape retention of the parts during debinding , the dimensional accuracy and other properties of the sintered parts. Hence, the binder is considered by many as the key to the success of a MIM process. Unfortunately, the development of binders has largely been empirical due to the lack of complete understanding of the underlying basic principles involved. Most binders were formulated based on "experience", rather than on scientific grounds.

Basic Binder Requirements

Table 1 lists the requirements of a binder system, which are further classified under different process considerations [1, 5]. The binder strongly influences the downstream operations and the final properties of the products. This inevitably leads to the numerous requirements. However, some of these attributes are contradictory to one another. Thus there is no perfect binder and the selection is dependent on the particular situation. Binders represent a compromise between various desired attributes.

Binder Classification

Binder compositions and debinding techniques are basically the main differences between various MIM/PIM processes. In fact, these two aspects are the main focus of many patents for MIM/PIM processes. Currently, there are various kinds of binders used in the MIM or PIM industry. Generally, binders can be broadly classified under the following groups:
- i.   thermoplastic based;
- ii.   thermoset based;
- iii.   gelation;
- iv.   freeze-forming;

Table 1. Binder Requirements [1, 5].

PROCESS CONSIDERATION	REQUIREMENT OF BINDER
Mixing, Viscosity and Uniformity of Mixture	. Extremely low viscosity . Good Adhesion to powders . Chemically passive with respect to powders
Molding	. Low viscosity at molding temperature . Low thermal expansion coefficient . High thermal conductivity
Green strength	. Good Mechanical Properties . Good Adhesion to powders
Uniform shrinkage	. No molecular orientation
Debinding	. Very low viscosity for wicking . Good solubility for solvent extraction . Easy pyrolysis for thermal debinding . No distortion, slumping or blow-out . Completely removable without any residue . Some presintering to retain shape after debinding . Decomposition temperature above mixing and molding temperatures
Economic	. Reasonable cost and available . Reasonable shelf life . Re-processable and reusable
Health	. Non-toxic, non-carcinogenic
Environmental	. Non-polluting

Thermoplastic/Wax Based

Thermoplastic based binders are by far the most widely used. These forms of binder usually contain a wax as a major component and a thermoplastic as the minor component. Additives are usually added for lubrication, viscosity control, wetting and improving powder-binder interaction. Debinding of such binders is normally achieved via thermal degradation, wicking, solvent extraction or even photo-degradation. Thermoplastics commonly used include polyethylene, polystyrene, polypropylene and ethylene vinyl acetate. Table 2 highlights some of the common thermoplastics used in PIM or MIM processes.

It is common practice to mix a number of thermoplastics, together with additives, to form a unique binder system that is tailored to a particular operation. This is due to the fact that a single component binder system was found to be incapable of satisfying the various requirements of a desired binder system. The choice of the components in the binder system depends on a number of factors; the green strength, the viscosity of the feedstock, the volume fraction of the powder and the method of debinding. The binder selection consideration for the primary component is indicated as before in Table 2. On the other hand, the secondary binder component is selected as according to the following requirements.

Table 2. List of common thermoplastics binder in MIM.

Types of Thermoplastic Binder	Investigating Organizations/Researchers	Literature/ Patents
Waxes (e.g. Paraffin, beeswax, camauba wax, polyethylene, Acrawax)	Curry; Bailey; Kiyota; Adee et al.	US Patent, 4,011,291,1977 US Patent, 3,285,873,1966 US Patent, 4,867,943,1989 US Patent, 4,225,345,1980
Polyoxyethylene/polyoxypropylene	Nagai et al.	US Patent, 4,898,902,1990
Polyamide	Ono et al.	US Patent, 5,002,988,1991
Organo-titinates	Storm	US Patent, 4,207,226,1980
Polyphenylene	Ohnsorg	US Patent, 4,144,207,1979
Glycols (e.g. polyethylene glycol, polypropane glycol)	Bailey; Wiech.Jr.	US Patent, 3,285,873,1966 US Patent, 4,197,118,1980
Polypropylene, polystyrene	Wiech.Jr.; Kiyota; Adee et al.; K.F.Hens,S.T.Lin, R.M.German, D.Lee	US Patent, 4,197,118,1980 US Patent, 5,006,164,1991 US Patent, 4,225,345,1980 JOM, Vol 41, No.8, August 1989, p. 19-24
Polyvinyl alcohol	Wiech,Jr.	US Patent, 4,197,118,1980
Polyvinyl acetate	K. Menke et al.	European patent, EP0428999,1990
Lactone resin	Nakanishi et al.	US Patent, 4,968,739,1990
Ethylene vinyl acetate (EVA)	B.O.Rhee, C.I.Chung (Rensselaer Polytechnic Inst.) R.Miura et al., (Research Center for Advanced Technologies) Renlund et al.	Proc. 4th APP Annual Meeting, MPIF,1990: JSW Technical Review, No.14 US Patent, 4,571,414,1986
Polybond (acrylic acid modified polypropytene)	B.O.Rhee, C.I.Chung (Rensselaer Polytechnic Inst)	Proc. 4th APP Annual Meeting, MPIF, 1990
Polymethy acrylate	Angermann.Yang, Van der Biest	Int.J. Powder Metallurgy, Vol 28, No.4,1992
Solid Polymer Solution	B.O.Rhee; (Rensselaer Polytechnic Inst.) G.B. Kupperblatt (Rensselaer Polytechnic Inst.)	Ph.D. thesis, RPI, 1992; M.Sc. thesis, RPI, 1990.

Table 3. Functions and requirements of the secondary component for thermoplastic binder system [5].

Function	Requirements
Viscosity reduction	- viscosity lower than the primary component - compatible with the primary component
Uniformity / Stability of the Mixture	- better adhesion to powder - better adhesion between components
Mold Release /Lubricant	- incompatibility with all binder components - viscosity lower than primary component
Green strength increase	- better adhesion to powders
Progressive debinding	- higher viscosity for wicking, less solubility for leaching, higher pyrolysis temperature for thermal debinding, or non-sublimating
Sintering/ Shape Retention / uniform shringkage	- high pyrolysis temperature for presintering, - no preferential orientation - complete pyrolysis without residue

Although there are many types of thermoplastic based binders being formulated, only a few binders are used in commercial production. The most popular are the wax-based binders. Wax is chosen as the major component because of its low viscosity, low melting point and low decomposition temperature. This would lead to ease of mixing, ease of molding and short debinding time. To further enhance the performance of the feedstock, a secondary component is added.

Thermoset Based

Thermosets, such as phenolics and epoxies, are polymers that hardened irreversibly by means of a cross-linking reaction. This usually happens at elevated temperature or upon mixing the resin with a hardener. Once the reaction is completed, a three dimensional cross-linked network structure is formed and the reaction is irreversible. Debinding is also accomplished by thermal degradation or solvent extraction. Table 4 shows some example of thermoset used in MIM or PIM.

Table 4. List of some common thermoset binders.

Types of Thermoplastic Binder	Investigating Organizations/Researcher	Literature/ Patents
Polycarbosilane	Su (GTE Laboratories Inc.)	US Patent, 4,939,197, 1990
Araldite resin	Strivens	US Patent, 2,939,199, 1960

Condensation crosslinking usually involves vapor formation as a by-product which would cause defects in the molded parts. Hence for MIM applications, only the addition crosslinking is of interest. The hardening or curing process is generally slow, so that the time needed to form a shape is long compared to that for thermoplastics. Moreover, the hardener needed for the curing process induce an additional component to the binder system which might cause mixing or compatibility problems. The advantage of using a thermoset binder is that it provides higher green strength due to the cross-linked structure. A natural compromise to this would be to mix a thermoplastic and a thermoset in one binder system. The thermoplastic serves to provide initial green strength while subsequent heating causes the thermosets to harden. As an example, Strivens in US Patent 2,939,199 dated 7 June 1960, used a thermoset, Araldite, mixed with a thermoplastic, Epok, and a wax to form a binder system for the injection molding of alumina [6]. Figure 3 shows the process flow by Strivens. Such an approach would help to shorten the molding time and retaining shape during debinding. However, the overall debinding time would probably be too long for such an approach to be economically feasible.

Gellation Method

Historically, investigators have recognized the limitations that the binder placed on the process. The polymeric binder system, which allowed the forming of complex shapes with particulates, is also the cause of many technical and economical problems. Hence, other alternative forms of binders are investigated. One of these is via the gelation approach.

Table 5. List of some common gelation type binders.

Type of Gelation Type Binder	Investigating Organizations/Researchers	Literature/Patents
Methyl Cellulose (aqueous-based)	Rivers, (Cabot Corp.); A.L. Salamone, J.A.Reed	US Patent, 4,113,480,1978 American Ceramic Society, Vol 58, No 6,1979
Silica Sol	Downing et al.; Blasch et al.	US Patent, 3,885,005 US Patent, 4,552,800
Agar, agarose	Fanelli et al. (Allied-Signal Research and Technology)	US Patent, 4,734,237,1988

Gelation is defined as the formation of a single molecule that extends via covalent bonding throughout the chemical structure. It is usually a hydration process. The massive molecule formed is a three-dimensional network which yields a high viscosity once it is formed. Low melting temperature liquids like water and alcohol may be trapped within the gel. The liquid, once evaporated due to elevated temperature, would result in a highly viscous structure that would bind the powder particles together. Debinding is achieved via evaporative or sublimative drying followed by

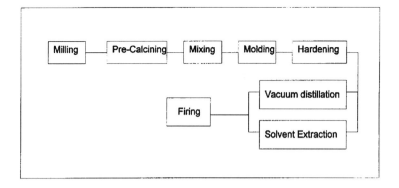

Figure 3. Thermoset binder approach for MIM process [6].

thermal degradation. Figure 4 illustrates an example of such an approach developed by Fanelli [7].

The gelation approach, as compared to the polymeric binder approach, has the following advantages.

i.   Both the processing temperature and pressure are lower. This leads to the use of lower capacity equipment and hence is very economical.

ii.   Debinding is faster as upon the initial rejection of fluid during gelling, there exists a network of porosity that facilitates the "bum-off" of the remaining binder during presintering.

iii. The required binder content is lower, about 2% to 4% as compared to 10% to 20% when using polymeric binders. This would mean a higher powder loading and hence reduced shrinkage after debinding and sintering.

However, there are drawbacks in using the gelation method. One drawback lies in the existence of a continuous liquid phase during thermal drying, which might lead to destructive capillary forces and result in rearrangement of the powders in the green parts during debinding. This would cause the internal homogeneity of the green parts to be disrupted. Another disadvantage is that the gelation process requires time and hence inevitably, slows down the molding process.

Freeze Forming

Freeze forming is another alternative technique aimed at overcoming the various problems faced by thermoplastics and thermosets. Figure 5 illustrates the process flowchart of a freeze forming process developed by Ceramics Process Systems [8, 9].

In a nutshell, the process consists of mixing the powder with a liquid which serves as a temporary vehicle. The liquid could be water or even organic compounds such as ethers and alcohols. Additives, like lubricants, cryoprotectants and dispersants, are usually added to the mixture for various purposes. The mixture, or slurry, is then injected under pressure into a closed die and concurrently the temperature is lowered until the mixture freezes and solidifies before ejecting out the part. The binder is removed by sublimative or evaporative drying. Sublimative drying or freeze drying, means drying at a pressure below the vapor pressure of the binder fluid and a temperature below the melting point of the fluid. On the other hand, evaporative drying would involve a formation of a continuous liquid phase during debinding.

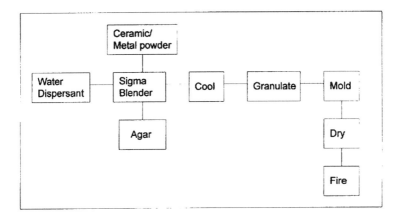

Figure 4. Gelation injection molding processing steps [7].

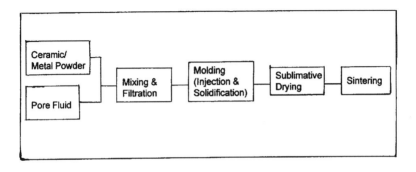

Figure 5. Process flowchart of an example of freeze forming
technique [8].

Table 6. Freeze forming type binder systems.

Type of Freeze-Forming Binder	Investigating Organization/Person	Article/US Patents
Water	Nesbit; Dennery et al.; Maxwell et al.; Weaver et al; Takahashi, (Mitsubishi Corp.) Occhionero et al., (Ceramics Process Systems)	US Patent, 2,765,512 US Patent, 3,567,520 US Patent, 2,893,102 US Patent, 4,341,725 US Patent, 4,965,027,1990 US Patent, 5,047,181,1991
Napthalene, Paradichlorobenzene, Camphor	Herrmann	US Patent, 3,330,892 & 3,234,308
Non-aqueous liquid (e.g. alcohols, alkanes, formic acid)	Sundback et al. , (Ceramics Process Systems)	US Patent, 5,047,182,1991

The advantage of using freeze drying is that there is very little shrinkage which can be controlled. In addition, some processes do not need distilled or deionized water nor pure or clean starting powders [10]. The dispersants added was claimed to be capable of overcoming the usual contaminants in tap water and in commercial ceramic or metal powders. This makes the process economical and robust. With the relatively higher freeze molding rate and shorter debinding time, the process is efficient and fast. With low and uniform shrinkage, parts' dimension tolerance is high. It is claimed that the final sintered part is capable of achieving above 97% densification. The dimensional tolerance is claimed to be less than $\pm0.5\%$ and the surface roughness as sintered is about 2.03 microns. The drawback of this approach is the need for modifications to existing molding equipment to incorporate the cooling agent (e.g. liquid nitrogen) as well as a special drying environment for debinding.

RHEOLOGICAL PROPERTIES

To minimize dimensional shrinkage and distortions after debinding, the concentration of powder in a feedstock is made very high, usually in the range of 50 to 80% by volume. During molding, the feedstock is essentially a highly concentrated suspension caused to flow and fill the mold cavity by the action of pressure and temperature. Throughout the MIM process, the feedstock is subjected to a wide variation of temperature, pressure and shear rate. Hence, it is important to understand the rheology of the feedstock under these conditions in order to optimize the process. Figure 6 shows the different shear rates experienced by a typical MIM feedstock.

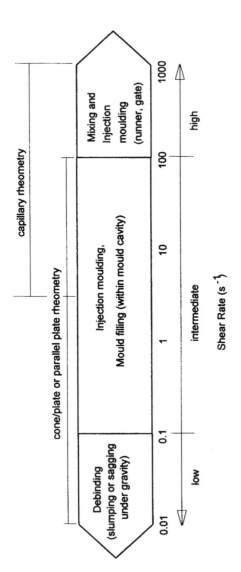

Figure 6. Range of shear rates experienced by a feedstock during a typical MIM process.

The entire range of shear rates can be divided into three regions, namely the low, the intermediate and the high shear rate regions. In the mixing and during injection molding process steps, the shear rate involved would fall into the intermediate and high shear rate regions. On the other hand, slumping behavior of the green parts during thermal debinding could be classified under the low shear rate regime. The rheological instruments commonly used in the rheological study of feedstocks include the cone-plate or parallel plate rheometers and the capillary rheometers. The former is normally used for the low to intermediate shear rate regimes while the latter is meant for the higher shear rate regions.

From literature, injection molding and mixing involves shear rates in the intermediate and high shear rate regions ($100\text{-}1000s^{-1}$) [1,11]. The rheology of feedstocks under such conditions form the main focus of most MIM or PIM research [11-17]. The capillary rheometer is usually used. However, the disadvantage in using this rheometer lies in the need to perform corrections for ends effects by either the Couette method or Bagley method [18].

The parallel plate or the cone-plate rheometers are capable of measuring the rheology of feedstocks at low shear rate regimes. Due to physical constraints, the shear rates investigated rarely go beyond 100 s^{-1}. However, the sensitivity and the ability to perform both the steady shear and oscillatory rheological measurements make this form of rheometers very attractive to researchers. The oscillatory measurement is a valuable tool for studying filled systems as a small amplitude could be used without disrupting the internal structure. The measured complex viscosity, η^*, could be related to steady shear viscosity, η , by means of a Cox-Merz rule [19]. The Cox-Merz rule states that a plot of complex viscosity with frequency is equal to a corresponding plot of viscosity with shear rate. That is

$$\eta^*(\omega) = \eta(\dot{\gamma})$$

However, while the Cox-Merz rule is valid for general unfilled and low concentration filled systems, it was discovered that the rule fails to hold for concentrated suspensions. It is important at this point to recognize that MIM or PIM feedstock is essentially a concentrated suspension at elevated temperature. A modified Cox-Merz rule was developed for concentrated suspensions by Doraiswamy et al [20]. The modified rule stated that the complex viscosity versus shear rate amplitude plot is equal to the corresponding viscosity versus shear rate plot. That is

$$\eta^*(\gamma.\omega) = \eta(\dot{\gamma})$$

Recent literature has shown that the modified Cox-Merz rule could fit experimental data very well [20-22]. Isayev and Fan [22] of University of Akron investigated both steady and oscillatory shear flow behavior of silicon-polypropylene ceramic compound. They reported that steady shear measurement with either parallel plate or capillary rheometers posed a number of problems. Difficulties encountered include melt fracture, flow instability, wall slip and melt stick/slip behavior. In such cases, it was recommended that oscillatory measurements could provide valuable information via the modified Cox-Merz rule.

Effects of Shear Rate

The effects of shear rate on the rheological behavior of feedstocks were extensively studied. Dependence of viscosity on shear rate can be expressed by the power law equation,

$$\tau = k \dot{\gamma}^{\,n} \tag{1}$$

where n is the exponent used to characterize the fluid. When n is unity, the flow is termed Newtonian; when n is less than unity, the flow is pseudoplastic or shear thinning and when n is larger than unity, the flow is considered dilatant or shear thickening.

Figure 7 shows a plot of viscosity versus shear rate for a MIM feedstock (58 vol. % carbonyl iron powder with EVA/beeswax binder) for 130, 150 and 170 °C. It can be observed that within the range of shear rate investigated, the feedstock exhibited shear thinning behavior.

Most MIM or PIM suspensions were found to exhibit first a Newtonian region, followed by a pseudoplastic region and finally, at high shear rates, a Newtonian region or a dilatant flow [24]. The pseudoplastic behavior was attributed to a combination of effects from the breaking up of agglomerates with increasing shear stress and the inherent pseudoplastic nature of the binder. It was recommended that a low shear sensitivity flow behavior at molding conditions ($100\text{-}1000\ s^{-1}$) would produce good molding behavior [13]. Dilatancy is generally encountered in suspensions at high shear rates [25, 26]. From literature, there exist a critical shear rate, which marks the onset of dilatant behavior [26, 27]. Dilatancy in concentrated suspensions has been attributed to rearrangement of the particles and collision of the particles [28]. Such effects cause greater inhomogeneity with increasing shear rates and hence, higher viscosity. Dilatant behavior is undesirable in injection molding and serve as a restriction to the processing parameters.

Yield Stress

Yield stress has been observed in MIM feedstock suspensions at low shear rates and low shear stress conditions [1,11,13]. A yield stress of a fluid can be defined as the limiting shear stress that has to be exceeded to initiate a shear flow. A fluid that exhibits such a yield stress is termed viscoplastic [29]. When an applied stress is less than a particular stress (yield stress), the

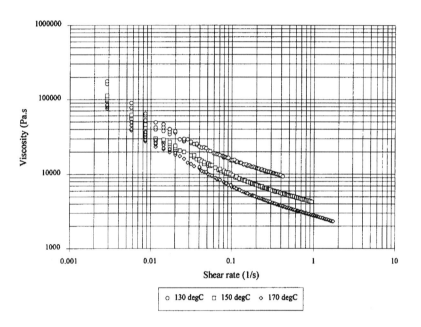

Figure 7. Viscosity versus shear rate plot for feedstock (58 vol.%
carbonyl iron powder with EVA/beeswax binder)
at 130, 150 and 170 °C [23].

fluid will not flow but deform plastically like a solid with a certain strain recovery upon the removal of the stress. When the stress exceeds the yield stress, the fluid will flow like a viscous fluid with a finite viscosity [29]. This concept is best illustrated by a plot of shear stress versus shear rate, as shown in Figure 8. The flow behavior of a normal viscous fluid (for example, polymer melt) would follow curve A. The curve would begin at the origin, indicating zero shear rate at zero shear stress. A viscoplastic fluid, on the other hand, would only start to flow after the shear stress has exceeded the yield stress, τ_y, as shown by Curve B.

Yield stress fluids can be found in a diversity of fields, such as biomedical, food processing and engineering materials. Important examples include concentrated suspensions, pastes, emulsions, gels, paints, blood and other body fluids.

In a concentrated suspension, the particles are in close proximity with one another. Interparticle interactions are present and cause a three dimension network to be formed. On the microscale, the yield stress can be viewed as the force per unit area necessary to overcome such interparticle interactions, and its magnitude is determined by the overall strength of the interparticle network. Lang and Rha [30] identified the factors that could affect the yield stress of a suspension and is listed in Table 7.

Table 7. Factors affecting yield stress.

I. STRENGTH OF PAIR INTERPARTICLE INTERACTION

A. Nature of interparticle operative force
B. Interparticle Distance
   a. Volume concentration of particles
   b. Net interparticle potential
   c. Packing configuration
C. Contact area/volume of particle
   a. Particle size and size distribution
   b. Particle shape
   c. Particle surface roughness

II. NUMBER OF PAIR INTERACTIONS PER UNIT AREA

A. Volume concentration of particles
B. Type of packing order of particles
   a. Particle shape
   b. Particle size distribution
   c. Packing density

The nature of interactions between particle could by attractive or repulsive, depending on the nature of the particles. Example of attractive forces include Van der Waals, hydrophobic bonding and hydrogen bonding while repulsive forces would include electrostatic and steric hindrance [30]. These interactions may act concurrently and would act only within a specific proximity. From this fact it could be deduced that there could exist a

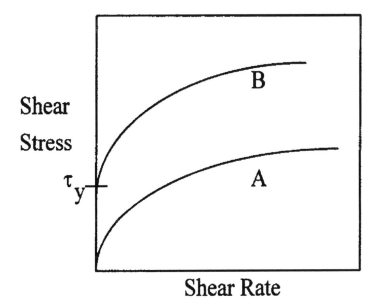

Figure 8. Yield stress of viscoplastic fluid.

minimum concentration of particles before an onset of yield stress would occur. Hence, at low concentrations, yield stress behavior would not happen at all. Despite a large amount of work being done on yield stress in concentrated suspensions, much of the mechanics of the yield behavior in suspensions still remains unanswered. There are even differing opinions on the existence of a true yield stress in fluids, that is, whether it is a real material property or just a virtual observation dependent on the time scale of the measuring techniques [31-35].

Measuring yield stress of concentrated suspensions can be carried out using various rheological techniques that can be broadly classified under two categories: the controlled rate rheometry and the controlled stress rheometry. A controlled rate rheometer deforms a specimen at a constant shear rate and measures the shear stress. On the other hand, a controlled stress rheometer imposes a constant shear stress on a specimen and then measures the corresponding strain. The latter approach involves a more sophisticated control system and is only introduced in the last ten years. These techniques can be further classified as direct (or static) or indirect methods (or dynamic). The indirect determination of yield stress involves the extrapolation of experimental shear stress - shear rate data to obtain a yield stress, which is the shear stress at zero shear rate. This is illustrated in Figure 9. It is evident that the choice of the model or methods yield differing values of yield stress.

The direct method involves measuring the yield stress directly and is independent of the shear stress - shear rate curves. However, the accuracy depends on the sensitivity of the equipment and the skills of the investigators. Reviews on the various measurement techniques of yield stress fluids have been described by Nguyen and Boger [29] and DeKee and Durning [33].Some of the techniques discussed include the vane method developed by Nguyen and Boger [29], parallel disk rheometry [36], parallel plate [37], cone and plate rheometry [38, 39], falling ball method [31] and the inclined plane method [29]. Windhab introduced a highly sensitive controlled stress capillary rheometer that was claimed to be capable of measuring viscosity at very low shear rates and yield stress of suspensions [40]. Of course, there is no single best method to measure yield stress of fluids. Every method has its advantages, as well as its limitations as highlighted in two recent reviews [29, 41]. Therefore, the choice of method would depend very much on the accuracy required and the intended application.

To date, there are various models developed to described the flow behavior of a fluid exhibiting yield stress. The three most common models used are listed as follows:

The Bingham Model [42]

$$\tau = \tau_y + \eta_p \dot{\gamma}, \qquad \tau \geq \tau_y \qquad (2)$$

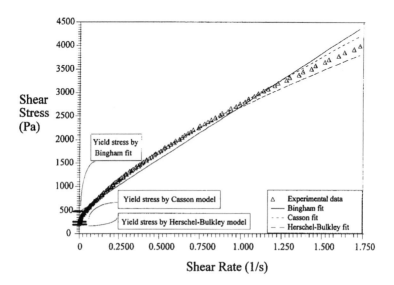

Figure 9. Determination of yield stress for feedstock (58 vol.% carbonyl iron powder with EVA/beeswax binder) at 170°C by different mathematical models [23].

The Herschel-Bulkley Model [43]

$$\tau = \tau_y + k(\dot{\gamma})^n, \qquad\qquad \tau \geq \tau_y, \quad k,n - constants \qquad (3)$$

The Casson Model [44]

$$\tau^{1/2} = \tau_y^{1/2} + (\eta_\infty \dot{\gamma})^{1/2}, \qquad\qquad \tau \geq \tau_y \qquad (4)$$

A simple set of constitutive equations was proposed by Doraiswamy et al [20] that described the behavior of a yield stress suspension. It incorporates a Hookean elastic response below yield stress and a power law behavior upon yielding. The equations are as follows.

$$\tau = G\hat{\gamma} \qquad |\hat{\gamma}| < \gamma_c, \quad \tau < \tau_y \qquad (5a)$$

$$\tau = \left(\frac{\tau}{|\dot{\gamma}|} + k|\dot{\gamma}|^{n-1} \right) \times \dot{\gamma} \qquad |\hat{\gamma}| < \gamma_c, \quad \tau < \tau_y \qquad (5b)$$

$\hat{\gamma}$ is a recoverable strain tensor.

k and n are power law parameters and G is the shear modulus. γ_c is the critical strain when the shear stress reaches yield stress. That is,

$$\tau = G\gamma_c \qquad (5c)$$

The recoverable strain behavior is described as

$$\frac{d\hat{\gamma}}{dt} = \dot{\gamma} \qquad |\hat{\gamma}| < \gamma_c, \quad \tau < \tau_y \qquad (6a)$$

$$\frac{d\hat{\gamma}}{dt} = 0 \qquad |\hat{\gamma}| = \gamma_c, \quad \tau \geq \tau_y \qquad (6b)$$

It can be observed that equation (5 a) described shear stress in terms of strain whereas equation (5b) in terms of shear strain rate. It can also be observed that equations (5) are but generalization of the Herschel-Bulkley model (3). There are more sophisticated constitutive equations and models developed but it will not be possible to cover them in this chapter. A detailed description could be found in a review by White [45].

A method of determining yield stress based on a combination of both the static and dynamic approaches was proposed by Foong et al [23]. It involves extrapolating to zero shear rate of the measured data obtained using a controlled stress rheometer over low shear rate regime. The obtained yield stress data was compared to those obtained using direct measurements and calculations with the Bingham, Casson and Herschel-Bulkley models. It was found that the proposed method obtained yield stresses lower than those obtained using direct measurements and Bingham model. Casson and Herschel-Bulkley models gave lower yield stresses but as these two approaches are indirect methods, the values are dependable on the shear thinning range of the flow data. The proposed method had the advantage of overcoming the limited resolution of a rheometer and the inconsistency associated with the use of mathematical models.

Effects of Solid Volume Fraction

Solid volume fraction or powder loading, ϕ , is the volume occupied by the solid particles as a fraction of the entire volume of the powder-binder mixture. The viscosity of a feedstock increases as more powder is mixed into the binder. The effects of solid volume fraction can be partially summarized and illustrated in Figure 10. If a small volume fraction of particles is suspended in a Newtonian liquid whose flow behavior is given by curve a, the viscosity of the suspension is uniformly raised to curve b. On further addition of particles, the viscosity is observed to continue to increase but becomes shear thinning, although Newtonian flow could still be possible at low shear rates (curve c). If however, the suspending medium is non-Newtonian, the flow behavior could follow curve c and addition of particles simply shifts the curve upwards to that of curve d. Further addition of particles, whether in Newtonian or non- Newtonian suspending medium, the appearance of an apparent yield stress occurs as shown in curve e. The slope of such a curve at low shear rates is found to have to value of-1. Finally, at solid volume fractions close to the critical value, shear thickening would occur, as shown in curve f.

To   aid in the study of the effects of solid volume fraction, a parameter known as the relative viscosity is used. Relative viscosity, η_r , is the ratio of the viscosity of the mixture to that of the pure binder. It was discovered that there exist a critical solid loading, ϕ_c, whereby the relative viscosity, η_r, becomes essentially infinite and the resulting mixture is too stiff to be considered viscous. The existence of a ϕ_c can be explained by the immobile fluid concept [1]. Figure 11 illustrates this concept with a three particle cluster. A minimum amount of fluid is required to fill the interstitial space. In addition, a certain amount of the organic binder would also be adsorbed onto the surface of the powder to form an immobile adsorbed layer [17]. This corresponds to a minimum amount of binder that is immobile and

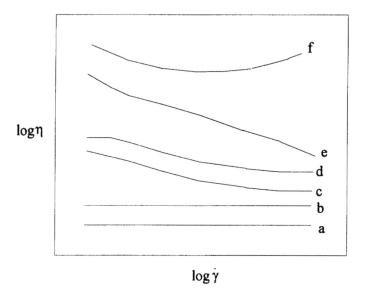

Figure 10. Qualitative illustration on the influence of increasing solid volume fraction on suspention flow behavior [26].

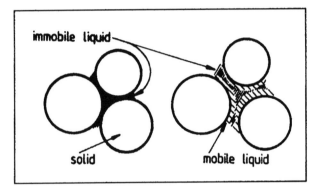

Figure 11. Two sketches of three particle cluster and the immobile fluid beetwen the particles when they are in contact. An excess of fluid (mobile) is needed to lubricate and initiate the flow process [1].

cannot participate in the flow process of the feedstock. At critical solid volume fraction, ϕ_c, the particles are essentially in contact and inhibited from rotating in shear flow. If the amount of binder is increased and the solid loading is decreased to ϕ, there will be an extra volume of binder which is equal to $(\phi_c-\phi)$ which is necessary to lubricate and allow shear flow. The amount of mobile fluid available will determine the viscosity of the feedstock. It can then be observed that the viscosity of the feedstock is dependent on the amount of mobile binder fluid in the mixture and not the total amount of binder fluid in the system.

The classical theory of Einstein describes a relationship between the solid volume fraction and the viscosity for monosized spherical particles dispersed randomly in a fluid [46]:

$$\eta = 1 + k_E \phi \tag{7}$$

The constant k_E is known as the Einstein coefficient and has the value of 2.5. However, this relation is applicable to only correlate loading of less than 15%. A number of expressions have since been developed to relate relative viscosity to solid volume fraction for concentrated suspensions. A typical characteristic of such equations incorporates the function of $(\phi_c-\phi)$ in the denominator. Therefore the relative viscosity can be correlated to the extra volume of binder over and above the immobile binder, $(1-\phi_c)$ as shown below:

$$\eta_r = 1 + k_E \phi \qquad \text{--Eilers[47]} \tag{8}$$

$$\eta_r = A \times exp\left(\frac{2.5\phi}{1 - k\phi} \right) \qquad \text{--Mooney[48]} \tag{9}$$

$$\eta_r = \left(\frac{\phi_c - 0.25\phi}{\phi_c - \phi} \right)^2 \qquad \text{--Chong et. al [17]} \tag{10}$$

It can be observed that Chong's equation approximates to Einstein's equation at very low solid loading where ϕ_c is 0.605 (for monosize spheres under body centered cubic packing). Stedman et al (16) measure the rheology of silicon nitride particulates/silicon carbide whiskers composite mixed with a polypropylene based binder. The relative viscosity data was found to fit Chong's equation well. Zhang and Evans [17] investigated on alumina-polypropylene suspensions and reported the limitations of Chong's equation. They modified Chong's equation and proposed a more general form by replacing the factor of 0.25 with a constant C. The experimental data was found to fit the modified form more closely.

$$\eta_r = \left(\frac{\phi_c - C\phi}{\phi_c - \phi} \right)^2 \tag{11}$$

However, it should be noted that all the above models must satisfy the following conditions [1]:

i.   The relative viscosity equals to 1.0 for the pure binder,

that is, $\lim_{\phi=0} [\eta_r] = 1.0$

ii.   From equation (3.6), the first derivative of the viscosity with solid content must equal to 2.5 in the limit of zero volume

fraction, that is, $\lim_{\phi=0} \left[\dfrac{d\eta_r}{d\phi} \right] = 2.5$

iii.   As the solid content increases, the relative viscosity must approach infinity at an asymptotic limit called the maximum

loading, ϕ_m , that is , $\lim_{\phi=\phi_m} [\eta_r] = \infty$

Effects of Temperature

Figure 12 shows the shear stress versus shear rate plot of a feedstock (58 vol% carbonyl iron powder with EVA/beeswax binder) at various temperatures. As the slope of the curves represents the viscosity of the feedstock, it can be observed that the viscosity decreases with temperature.

It has been well documented in literature that the viscosity of feedstocks decrease with increasing temperature. However, the effects of temperature on feedstock is not always as simple as that of a single component binder. Kupperblatt [49] reported a complicated temperature dependence of binder viscosity which was caused by different melting points of the various components of the binder. This may result in situations whereby some components of a binder remain semi-solid while others are in liquid form. Moreover, the effects of temperature is magnified by thermal expansion of the powders and the binder. Thermal expansion of the binder is generally higher than that of the powders. Consequently, the effective solid volume fraction of the mixture actually decrease with an increase in temperature causing the viscosity to drop even further. Hence, with an

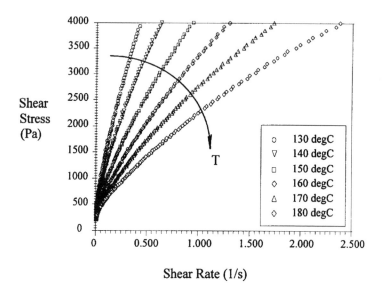

Figure 12. Shear stress versus shear rate plot for feedstock (58 vol.% carbonyl iron powder with EVA/beeswax binder) at various temperature [23].

increase in temperature, the relative rate of viscosity drop is typically faster in the feedstock than that in a pure binder [1]. At temperatures far above the melting point of a given binder, the temperature dependence of viscosity follows an Arrhenius type equation,

$$\eta_{(T)} = A \times exp\,(E_a/RT) \qquad\qquad (12)$$

where $\eta_{(T)}$ is the viscosity at a constant shear rate, A is an empirical constant, E_a is the activation energy for shear viscosity, R is the gas constant and T is the absolute temperature. E_a is a measure of the shear viscosity's sensitivity towards temperature change. A high E_a would indicate high sensitivity towards temperature change. Cao et al [15] investigated the injection molding of iron powder feedstocks with three different binder formulations, namely LMW-PP, EVA and HDPE based. It was reported that E_a of the feedstocks is dependent on the molecular weight of the polymer and the binder-powder interaction. Moreover, it was claimed that a low E_a in the range of the molding conditions would give good molding characteristics. E_a can be determined from the slope of a plot of $log(\eta)$ versus (1/T) at a reference shear rate. This is known as the Arrhenius plot. An Arrhenius plot for the same feedstock shown in Figure 12 is plotted in Figure 13. The data was obtained at a reference shear rate of $0.1s^{-1}$ A line was fitted by means of least square method and the slope is determined. E_a could be obtained by multiplying the slope of the fitted line with the universal gas constant (8.3144 J/mol.K).

Foong   et al [23] showed that yield stress is dependent on temperature. Figure 14 shows the variation of yield stress with temperature for three carbonyl iron powder   feedstocks. Feedstocks No. 2, 3 and 4 contained binder with EVA/beeswax   ratio of 40/60, 60/40 and 70/30 respectively. It can be observed that yield stress decreases non-linearly with increasing temperature. This observation contrasted the views of Malkin [50]. He proposed that the yield stress of a filled polymer melt is independent of temperature. However, as there were neither actual experimental data nor references quoted, the proposal was difficult to be verified.

At   this stage, the reason for the decrease in yield stress with temperature is not fully understood. However, Tanaka and White [51, 52] showed the dependence of yield stress on the solid volume fraction of fillers in a concentrated suspension using constitutive models. Experimental data were also presented to further support this dependence. It was reported that yield stress increases with increasing solid volume fraction. As described in the previous section, Lang and Rha [30] who listed the effect of the volume fraction on yield stress, further supported this notion. In our case, the solid volume fraction of the feedstocks was computed based on density data at room temperature. However, due to a difference in thermal expansion coefficients of the binder and the powder, the effective solid volume fraction changes. In the event of an increase in temperature, the binder, in a liquid

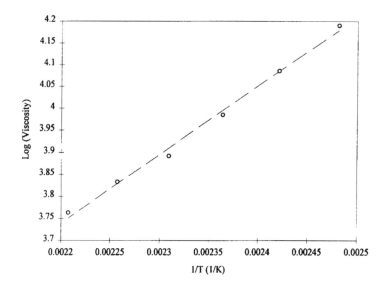

Figure 13. Arrhenius plot for feedstock (58 vol.% carbonyl iron powder with EVA/beeswax binder) at $0.1s^{-1}$ shear rate.

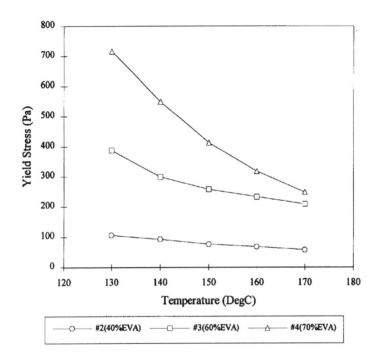

Figure 14. Yield stress versus temperature plot of feedstocks Nos. 2, 3 and 4 [23].

phase, probably expands more than the powder. This leads to a decrease in the effective solid volume fraction, which in turn causes a decrease in yield stress of the feedstock. Nevertheless, this is just one possible explanation for the decrease of yield stress with temperature.

The dependence of yield stress on temperature can be correlated by an Arrhenius type relationship similar to equation (7). The expression is as follows [23].

$$\tau_{(y)} = A \times exp\,(E_y/RT) \tag{13}$$

τ_y is the yield stress, A is a pro-exponential function, E is the activation energy for yield stress, R is the universal gas constant (8.3144 J/mol.K) and T is the absolute temperature. The activation energy for yield stress, E_y, which is similar in concept to activation energy for shear viscosity, gives a measure of the sensitivity of yield stress to thermal energy. A linear relationship as described by equation (8) can be obtained by plotting log (τ_y) versus (1/T) as shown in Figure 15. To determine E_y, the gradients of the plots were determined and multiplied by the universal gas constant, R.

Effects of Binder Composition

A low molecular weight component is usually added to the binder system as a plasticizer to reduce the viscosity of the feedstock. Usually this component is in the form of a wax. Beeswax, paraffin wax, microcrystalline wax and camauba wax are the common types of wax used in MIM.

Figure 16 shows a plot of viscosity versus shear rate data for carbonyl iron powder feedstocks with different binder composition. The binders are composed of EVA/beeswax in proportions of 40/60, 60/40 and 70/30. It can be observed that as the EVA content in the binder is increased, the viscosity is increased accordingly.

From Figure 14, it can be observed that yield stress increases with an increase in EVA concentration in the binder. One possible explanation for such a phenomenon is by assuming the formation of an immobile absorbed layer of binder molecular chains on the iron particles surface. The formation of such an interface or mesophase layer [53] would effectively increase the apparent size of the particles and in turn increase the effective solid volume fraction of the feedstock [54]. The increase in effective solid volume fraction would in turn lead to higher suspension yield stress by the same reasoning described in the previous section [51, 52]. The thickness of this absorbed layer corresponds to the random coil dimension for the molecular chains. The chain end-to-end distance, h, of a polymer molecular chain ranges from 20 to 100 nm and is given by the following expression [55].

Log (Yield Stress)

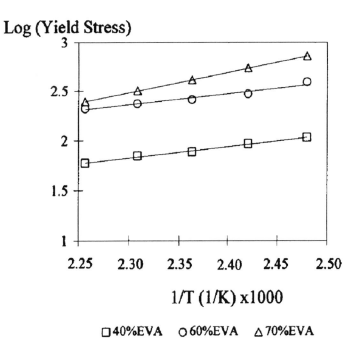

Figure 15. Plot of log(yield stress) versus (1/T) for feedstock No. 2 (40%EVA), No. 3 (60%EVA) and No. 4 (70%EVA). [23].

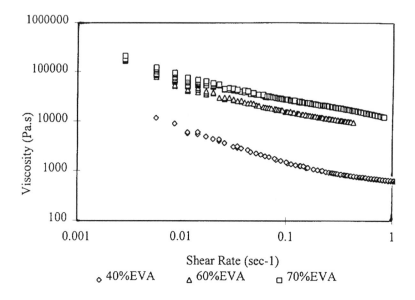

Figure 16. Viscosity versus shear rate for carbonyl iron feedstocks of different binder composition.

$$h = 3l\sqrt{n} \qquad\qquad (14)$$

where l is the bond length and n is the number of links per chain. Polymers adsorbed on high energy surfaces leaving loops and tails rather than as a flat molecular layer [56] and the thickness of the absorbed layer is, therefore, comparable to the value of h. Figure 17 illustrates the concept of the absorbed layer by means of the Hashin model [53].

Referring to Figure 17, the thickness (r_i - r_f) would be a function of h. With the absorbed layer taken into account, the effective solid volume fraction, ϕ_c, would be [17, 57],

$$\phi_c = \phi(1 + k\rho Sh) \qquad\qquad (15)$$

is the solid volume fraction, ρ is the density of the iron particles, S is the specific surface area of the particles and k is a constant ($0 < k < 1$) which defines the fraction of the absorbate that is sufficiently immobile to be considered as a solid phase.

With the concept of an absorbed layer of immobile molecules, it can then be used to explain the effects of EVA concentration in the binder on the yield stress of the feedstock. Firstly, due to polarity of the acetoxy groups on the EVA molecules, the EVA molecules would probably have a higher affinity towards the iron particles. Hence, it can be assumed the EVA molecules would be absorbed more readily onto the particle surface than the beeswax molecules. An increase in EVA concentration would greatly promote the formation of the absorbed layers, therefore, increasing ϕ_c and leading to a higher yield stress. Another reason could be that because the dimension of the absorbed layer is dependent on the molecular weight of the absorbed molecules as shown in equation (9) (h is a function of n which is proportional to the molecular weight of the polymer). With EVA molecules having a relatively higher molecular weight, the thickness of the absorbed layer would be higher compared to that with beeswax molecules. Hence, the increase in ϕ_c due to EVA would be more substantial than the wax. Based on this argument, it can be rationalized that an increase in EVA concentration in the binder would eventually lead to a higher yield stress.

Besides viscosity and yield stress, the binder composition affects very significantly the thermal sensitivity of the feedstock's shear viscosity and yield stress. Figures 18 and 19 show, respectively, the carbonyl iron powder feedstock's activation energy for shear viscosity and yield stress with respect to EVA content in the binder.

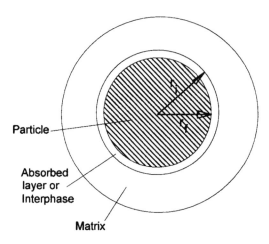

Figure 17. Model of a particle with absorbed layer of matrix where r_f and r_i represent the radii of the particle and that the absorbed layer taken into consideration.

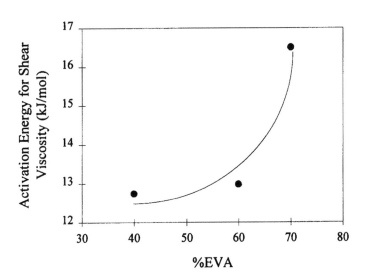

Figure 18. Variation of activation energy for shear viscosity with EVA content of carbonyl iron powder feedstock with different binder composition.

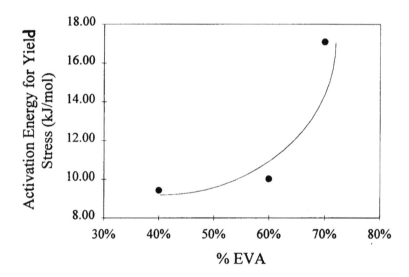

Figure 19. Variation of activation energy for yield stress with EVA content of carbonyl iron powder feedstock with different binder composition.

In comparing Figures 18 and 19, it can be observed that there exists a striking resemblance between the two plots. The effect of EVA concentration in the binder on the feedstock's E_a, activation energy for shear viscosity, is observed to be very similar to that on E_y, activation energy for yield stress. This observation seems to suggest a close relation between the two parameters. This is turn suggests that the mechanisms involved for both thermal sensitivity of shear viscosity and yield stress could be similar.

A probable explanation on the effect of EVA concentration on E_a and E_y could be achieved through the concept of the absorbed layer of immobile binder molecules on the particle surface. As was previously explained, the absorbed layer cause an increase in ϕ_c. However, it should be emphasized that this would require the absorbed molecules to be sufficiently immobile to be considered a solid phase. With an increase in temperature, the increase in the level of thermal energy would enable the molecules to gain enough kinetic energy to be more mobile. Consequently, the proportion of immobile molecules in the absorbed layer would fall. The value of k in equation (10) would correspondingly decreased. This eventually would cause ϕ_c to fall and increase interparticle distance. Hence, the yield stress or shear viscosity would start to decrease.

With this reasoning, it would follow that the increase in yield stress or shear viscosity due to the immobile absorbed layer is highly sensitive to temperature change. With a higher EVA concentration in the binder, larger amount of EVA would be absorbed on the particle surface. Although this has an effect of increasing the yield stress or shear viscosity of the feedstock, it would also render these properties more sensitive to temperature changes. Hence, an increase in the amount of EVA in the binder would raise the E_a or E_y.

THERMAL PROPERTIES

Throughout the entire MIM process, the feedstock is repeatedly subjected to elevated temperatures. Hence, it is important to understand the thermal properties of feedstock in the formulations of optimal processing parameters and aiding the development of simulation packages for the process. This section highlights some important thermal properties which include expansion coefficient, change in solid volume fraction, thermal conductivity, softening/melting points and thermal decomposition behavior.

Residual Stress and Thermal Expansion

During injection molding, pressure is required to ensure that the feedstock completely fills the mold cavity. When the mold is completely filled, the compact is cooled in the mold under pressure. The pressure is only removed at the end of the molding cycle. Residual thermal stress is always present in conventional thermoplastic injection molding due to macromolecule orientation that is "frozen in" during rapid cooling of the molded part. This residual stress could cause shape distortion and warpage after molding [58]. However in MIM (i.e., highly filled polymer systems) due to the vast difference in thermal expansion coefficients of the powder and the binder, coupled with cooling under pressurized condition, the internal stress in the MIM molded part has an additional component. This stress is due to difference in thermal expansion coefficient which can be estimated by the expression [1, 59]:

$$\sigma = E_m \times \Delta T \times (\alpha_m - \alpha_f) \tag{16}$$

where σ is the stress, E_m is the elastic modulus for the matrix (binder), ΔT is the temperature change, α_m and α_f are the thermal expansion coefficients of the matrix (binder) and the filler (powder) respectively. Upon cooling from processing temperature, the thermal contraction of the binder is usually higher than that of the powder, hence the net effect is that such a stress is compressive in nature acting, on the powder particles [59, 60]. This imposes a squeezing force on the particles. Hence, even if the adhesion between the binder and the particle is poor, there might not be any relative motion across the interface due to the large compressive stresses and resulting friction. However, in regions where the particles are in contact, especially at high powder loading, the thermal shrinkage of the matrix binder can lead to local regions of tensile rather than compressive stress. In such cases, poor adhesion would allow debonding at the interface and result in a cavity or pore. In severe cases, the result is equivalent to that of a foam if the particles are free to move inside the cavity [59].

The rule of mixture can be used to estimate the effective thermal expansion coefficient of the feedstock [1].

$$\alpha = \phi \alpha_f + (1 - \phi) \alpha_m \tag{17}$$

where ϕ is the solid volume fraction, α_f and α_m are the thermal expansion coefficients of the powder and binder respectively. However, due to the vast difference in the thermal expansion of the powder and binder, the solid volume fraction would change in the event of a change in temperature. The change in powder loading can be calculated using the following expression [1,61]:

$$\phi_{TO} = 1 + \left(\frac{1 - \phi_{TI}}{\phi_{TI}} \right) \times \left(\frac{1 + \alpha_m (T_0 - T_1)}{1 + \alpha_f (T_0 - T_1)} \right)^3 \tag{18}$$

where T_1 and T_0 are the temperatures of the final and initial conditions, (e.g. T_1 is the ambient temperature and T_0 the processing temperature). Equation

(18) would enable the estimation of the actual solid volume fraction during processing (e.g. mold filling) with an initial solid volume fraction estimated at ambient conditions. This equation would be useful in the analysis of the processing of the feedstocks, as well as for simulation of the process.

Thermal Conductivity

A simple approach to estimate the effective thermal conductivity is to use the rule of mixture.

$$\kappa = \phi \kappa_f + (1 - \phi) \kappa_m \tag{19}$$

where κ_f and κ_m are the thermal conductivity of the feedstock and the matrix (binder) respectively. ϕ is the solid loading of the particles. However, such an expression would require the particles to be in contact with one another in the heat flow direction (similar to the long fibre composites) as discussed by Rhee [13]. An alternative expression used by Rhee is the Jeffrey's equation [13]:

$$\frac{\kappa}{\kappa_m} = 1 + 3\zeta\phi + \phi^2 \left(3\zeta^2 + \frac{3\zeta^3}{4} + \frac{9\zeta^3}{16} \frac{\lambda + 2}{2\lambda + 3} + \frac{3\zeta^4}{2^6} \right) \tag{20}$$

where $\lambda = \kappa_f / \kappa_m$ and $\zeta = (\lambda - 1) / (\lambda + 2)$.

The thermal conductivity of the feedstock can also be approximated by Nielsen's equation [Nielsen (1973)]:

$$\kappa = \kappa_m \frac{1 + AB\phi}{1 - BC\phi} \tag{21a}$$

$$A = k_E - 1 \tag{21b}$$

$$B = \frac{(\kappa_f / \kappa_m) - 1}{(\kappa_f / \kappa_m) + A} \tag{21c}$$

$$C = 1 + \frac{1 - \phi_m}{\phi_m^2} \phi \tag{21d}$$

ϕ_m, the maximum packing volume fraction, is defined as the ratio of the true volume of the particles to the actual volume occupied by the particles. It is related to the packing efficiency of the particles and hence is a function of the powder characteristics, k_E is the Einstein coefficient but it has the value of 3.0 for dispersed spheres. Values of ϕ_m and k_E for different systems are listed in [63]. Due to the high thermal conductivity of the powders, the feedstock normally possess higher conductivity than the pure binders. A high thermal conductivity of the feedstock ensures heat is transferred fast enough to the entire compact. This will ensure a more even temperature gradient in the compact, minimizing thermal stress and hence minimizing shrinkage cracks. It also means that cooling is fast enough for higher molding cycles. However, the high conductivity also causes potential problems such as premature gate freezing [13].

Thermal Decomposition Behavior

During thermal debinding, the feedstocks are heated until the binder is removed via thermal decomposition. Hence, it is important to study the thermal decomposition of the binder in the presence of the powder. The TGA (thermogravimetric analysis) is usually used for this study. The samples would be heated at programmed heating rate with a stream of nitrogen gas purging the furnace's interior to reduce oxidation of the metal powder in the feedstocks.

Figure 20 shows TGA plots for carbonyl iron powder feedstock samples of F46030, F46040, F46050, F46060. The four feedstocks have different EVA/wax composition in the binders. The powder loading were kept constant at 58 vol.%. The TGA plots exhibited a three stage decomposition. In the first stage, the beeswax started to vaporize at about 200 °C. The weight loss accelerated to a maximum at about 330 °C as the beeswax, as well as the VA group in the EVA began to degrade. Weight loss then decelerated as it approached 400 °C, towards the second stage which began at about 420 °C. At this stage, the polyethylene backbone of the EVA started to decompose. The weight loss decelerated again as it approached 500 °C. The third stage decomposition started at about 550 °C was due mainly to the partial decomposition of the carbonyl iron powder. Hence, the amount of powder in the feedstock samples can be estimated from the TGA plots at the end of the second stage. From the plots, the amount of powder in the feedstock samples was found to constitute approximately 92 wt.% (about 58 vol.%).

Softening Temperature

The softening temperature is another alternative check for compatibility between the components of the binder, as well as that between the binder and the powder. Softening temperature of a mixture can be measured by a Thermo-Mechanical Analyzer (TMA). When a secondary

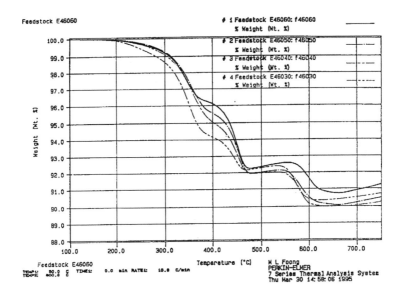

Figure 20. TGA plots of feedstock samples of F46030, F46040, F46050 and F46060.

component (e.g. a polymer) is compatible with the major component (e.g. a wax), the softening temperature of the binder mixture will be higher than that of the latter. On the contrary, if the polymer is not soluble in the wax and forms separate entities in the wax matrix, the softening temperature of the binder will be close to that of the wax. Figure 21 shows the softening points of paraffin wax, various polymers (secondary components), various binder mixtures and the feedstocks made from the binder and a fixed concentration of an iron powder.

The abbreviations are as follows:

PW   = paraffin wax,
EVA-A   = ethylene vinyl acetate
PB   = polybond
HDPE   = high density polyethylene
LMW-PP   = low molecular weight polypropylene

It can be observed that the softening points of the binders with HDPE and EVA are closer to the polymer components. Hence, these two polymers are more compatible with the PW than the rest of the polymers. It can also be observed that for EVA and PB, the feedstock's softening points are higher than that of the pure binders. This can be attributed to good adhesion between the binder and the powder. Both EVA and PB have molecules polar groups and hence, have greater adhesion to the powder particles.

The softening temperature increases as the powder loading is increased. Figure 22 shows a plot of softening temperatures for carbonyl iron powder feedstocks (with EVA/beeswax binder) of varying powder loading. The softening temperature can be observed to be increasing steadily with powder loading.

Melting and Recrystallisation Temperature

The melting and recrystallisation behavior of a binder blend is strongly affected by the degree of interaction between the constituents and the morphology of the resulting blend. A study of these behavior would aid in the understanding the interactions between the components in the binder. The compatibility and interaction between the binder components is important for preventing binder separation and ensuring stable homogeneity.

From literature, the melting and recrystallisation temperatures of a semi-crystalline polymer is depressed in the presence of a polymeric diluent and is governed by the equation [64],

$$\frac{1}{T_m} - \frac{1}{T_M^o} = - \frac{R\upsilon_c}{\Delta H_u \upsilon_a} \phi_a^2 \chi \qquad (22)$$

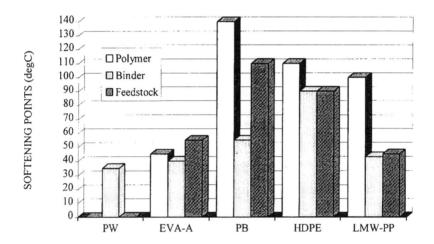

Figure 21. Softening points of paraffin wax, polymer components, wax-based binders and feedstocks [13].

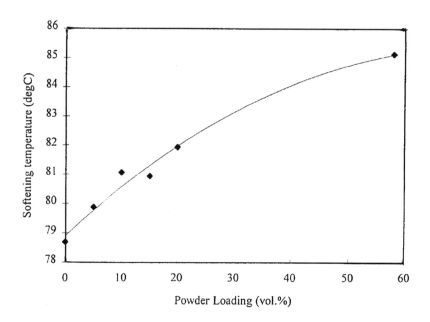

Figure 22.   Effect of powder loading on the softening temperatures of feedstocks.

T_m and $T_m^{\,\circ}$ are equilibrium melting temperatures of the polymer blend and the pure polymer respectively, υ_c and υ_a are the molar volumes of the repeat units of crystalline and amorphous polymer, respectively; ΔH_u is the enthalpy of fusion of the perfect crystal; ϕ_a is the volume fraction of the amorphous polymer; and χ is the polymer-polymer interaction parameter and is dependent on the heat of mixing but independent of combined entropy of mixing. Hence, melting temperature depression is only dependent on the degree of interaction between the polymer, provided the samples are crystallized and melted in the same manner [65]. To determine the interaction parameter, an accurate measurement of the equilibrium melting temperature is needed. However, for comparison purpose the non-equilibrium melting temperature depression can be used to yield a qualitative, rather than a quantitative, estimation of the level of interaction between the components.

The DSC (differential scanning calorimetry) is used for investigating such behavior. Figure 23 shows a DSC output for a binder sample containing EVA/beeswax of ratio 40/60. Both the melting and recrystallisation curve show a twin peak plot. The first peak (on the left)represented the melting or recrystallisation of the beeswax in the binder while the second peak (on the right) represented that of the EVA copolymer. The area under the peak, indicated by AH, would represent the energy consumed or evolved during the melting or recrystallisation process.

MECHANICAL PROPERTIES

MIM feedstocks can be viewed as particulate filled polymers with very high powder loading. These feedstocks are normally fragile and brittle due to high powder loading. An understanding of the mechanical properties of the feedstock is important to binder formulation. Poor mechanical properties causes difficulties in handling the molded parts after molding and in severe cases cracks after molding. Hence, the binder must be formulated as such to increase the mechanical strength and modulus.

In the study of mechanical properties of particulate filled polymers, numerous models were developed to predict the effect of the particles on tensile or shear modulus. Most of these were derived from rheological models such as Einstein's, Eilers and Mooney's equations. A strong relationship exists between rheology and mechanical properties measurements and such correlations were studied by Gahleitner et al [66], as well as by Pukansky and Tudos [67]. There seems to have a direct relation between viscosity and shear modulus [59]. However, compensation has to be taken for matrix's Poisson ratio which is lower than 0.5 as shown by Nielsen and Landel [59]. Nevertheless, these equations on modulus predictions can be broadly classified under two groups.

The first group include only constants of the matrix and the particles.

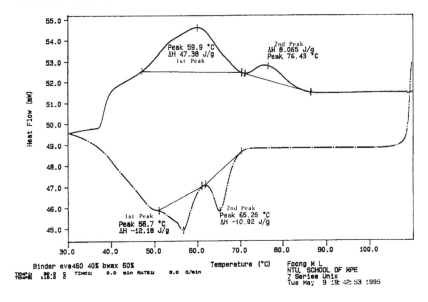

Figure 23.   DSC output for binder containing 40/60 ratio
of EVA/beeswax.

There are no adjustable parameters. Such models include the Kemer equation [68], Halpin-Tsai equation [69] and Chow equation [70]. The second group of equations, on the other hand, incorporate adjustable parameters to account for interactions between particles as well as between the matrix and the particles. Factors such as critical solid volume fraction, degree of agglomeration and powder-matrix adhesion are taken into account. Equations and models under this group would include the Nielsen generalized equation [71] and the modified Kemer equation [72,73].

Much work had been done and research is still in progress in the field of mechanical properties of particulate filled composite materials. It would not be possible to include all these information in this section. Nevertheless, some of the more important models and concepts, with emphasis on the interaction between the matrix(binder) and particles shall be reviewed.

The Kemer equation [68] can be used if the particles are near spherical and there is good adhesion between the particles and the matrix. For particles that are much more rigid than the polymer matrix, the equation can be expressed (up to moderate powder loading) as follows:

$$\frac{G}{G_m} = 1 + \frac{15(1-\nu_m)}{8-10\nu_m} \frac{\phi}{(1-\phi)} \qquad (23)$$

where ϕ is the solid volume fraction, ν_m is the Poisson's ratio of the matrix while G and G_m are shear moduli for the composite and matrix respectively. Halpin and Tsai [69] showed that the Kemer equation and many other equations can be generalized as follows:

$$\frac{M}{M_m} = \frac{1-AB\phi}{1-B\phi} \qquad (24a)$$

In this equation, M and M_m are the elastic moduli (shear, Young's or bulk) of the composite and matrix respectively, constant A takes into account such factors like aspect ratio of the particles, Poisson's ratio of the matrix, and constant B is defined as follows:

$$B = \frac{(M_f/M_m)-1}{(M_f/M_m)+A} \qquad (24b)$$

where the subscripts f and m refers to the particle and matrix phases respectively.

Lewis and Nielsen [71] showed that the Halpin-Tsai equation could be further generalized as follows:

$$\frac{M}{M_m} = \frac{1 - AB\phi}{1 - B\psi\phi}$$

(25a)

where $A = k_E - 1$

(25b)

$$B = \frac{(M_f/M_m) - 1}{(M_f/M_m) + A}$$

(25c)

$$\psi \approx 1 + \frac{1 - \phi_c}{\phi_c^2}\phi$$

(25d)

It can be observed that the constant B is basically the same as that for Halpin-Tsai equation. Constant A is related to the generalized Einstein coefficient k_E. The value of k_E is not a constant but depends on the state of agglomeration of the particles. It has the value of 2.5 for perfect dispersion of spherical particles and perfect matrix-particle adhesion. The value will decrease if there is dewetting and slippage at the matrix-particle interface but will increase if there is agglomeration [74]. However, k_E is applicable only to matrix with Poisson's ratio of 0.5. For other Poisson ratios, a conversion is required and can be done easily with a list of relative Einstein coefficients for various Poisson's ratios presented in the works of Nielsen and co-workers [59, 63, 71]. Finally, ψ depends on the maximum or critical packing fraction of the filler in the matrix.

Mukhopadhyay et al. uses a modified Kemer equation to evaluate the dynamic mechanical properties of silica filled ethylene vinyl acetate copolymers. The expression for the relative storage modulus is as follows [72, 73]:

$$\frac{E_f'}{E_m'} = \frac{1 + AB\phi}{1 - B\phi}$$

(26a)

where $A = k_E - 1$ and k_E has the value of 2.5 (dispersed spheres) while constant B is a matrix-particle interaction term. It was further found that the relative loss modulus of the system followed the following form:

$$E_f''/E_m'' = (1 - B\phi)^{-1}$$

(26b)

where B is the same term in (21a). It was found to agree with their data, however, the experiments were limited to solid volume fractions below 0.3.

Validity of equation (21b) for higher powder loading is uncertain. Ziegel and Romanov [72] proposed a more convenient empirical equation as shown below:

$$E_c^{''} / E_m^{''} = [\, 1 - (B\phi)^n \,]^{-1} \qquad (26c)$$

A plot of logarithm of $(1\text{-}E^{''}_c/E^{''}_m)$ versus logarithm of ϕ would yield the exponent n as the slope. The parameter B can also be evaluated from the intercept. It was shown in the same report that the parameter B obtained from equation (21c) agreed well with those used to fit the modified Kemer equation (21a) to the experimental relative storage modulus data.

Figure 24 shows a plot of relative storage modulus versus volume fraction of iron powder in the feedstock. The relative storage modulus obtained at various dynamic stress frequency are shown. The data obtained at various frequency were very close and seems to be independent of the frequency. Both the Nielsen equation and modified Kerner equation were used in an attempt to fit the experimental data. The modified Kerner equation fits the experimental data at low powder loading very well. However, it failed to predict the data at powder volume fraction of 0.58. In contrast, the Nielsen equation fits the data for the entire range of volume fraction.

The stress-strain behavior of a polymer is usually dramatically changed by the addition of particulate fillers. Generally, fillers cause a drop in the elongation to failure, increase the elastic modulus and often induce a yield point on in the stress-strain curves of ductile polymers and elastomers. Figure 25 shows schematically the effect of addition of fillers to a ductile polymer or elastomer.

The two basic deformation mechanisms in a polymer are by shear yielding and crazing. In a composite system, such as the MIM feedstock, an additional mechanism, dewetting, is observed [75]. The yield point is actually due to crazing effect or a dewetting effect in which the particle-matrix interface is destroyed and resulting in a dramatic drop in the modulus of the material [59].

In analyzing the micromechanics of dewetting and interfacial strength in particulate filled polymers, Zhuk et al [60] proposed model as shown schematically in Figure 26. The dewetting process was observed under an optical microscope and video recorder. It was observed that the dewetting would progress until a unique debonding angle before fracture of the composite occurs. The debonding angle is characteristic of a particular particle-matrix system. The extent of debonding angle could hence indicate the degree of particle-matrix adhesion, 0° for perfect adhesion and 90° for zero adhesion. It was found that the measure debonding angles were close to 68° for the glass bead-PP or PE systems and agree well with the proposed model based on energy Griffith theory. It was found that residual thermal

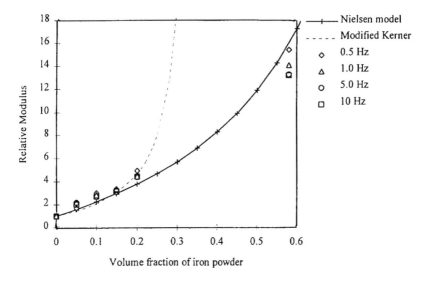

Figure 24.   Plot of Nielsen equation and modified Kerner equation with experimental relative storage modulus data obtained at various dynamic stress frequency.

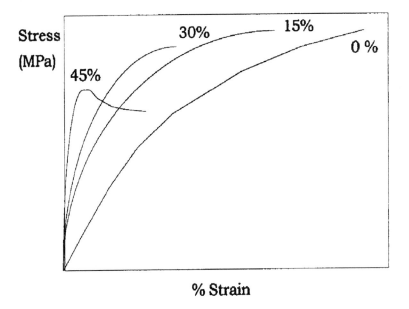

Figure 25.   Schematic stress-strain curves of a ductile polymer filled with rigid particulates. The numbers on the curves represent the solid volume fraction [59].

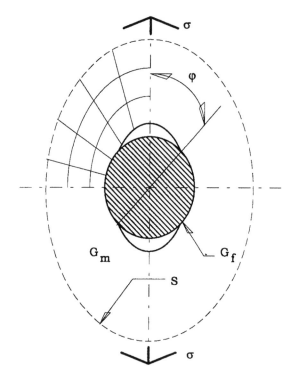

Figure 26.   Schematic representation of a deformed shell, S, of matrix
stress, σ. Partial dewetting occurs with a debonding angle
of φ [60].

contraction stress would lead to a decrease in the debonding angle and interfacial friction suppresses shear mechanism of crack propagation.

In the dewetting process, voids or cavities are created at the poles of the particle as shown in Figure 26. This leads to specimen dilation (increase in volume). Fracture then follows by the usual crazing mechanism of the matrix.

From this point, it can be seen that yielding or tensile strength is dependent on particle-matrix adhesion, specific surface area of the particles and thus the volume fraction. The dewetting process also causes the effective load bearing area to be reduced. Based on these factors, Pukanszky et al proposed an expression to describe the tensile yield stress and tensile strength [75].

$$\sigma_T = \sigma_{Tm} \lambda^n \frac{1 - \phi}{1 + 2.5\phi} \, exp\,(B\phi) \tag{27}$$

σ_T and σ_{Tm} are the true tensile strength of the composite and the matrix respectively, λ is the relative elongation, n characterizes the strain hardening effect of the matrix and B is a parameter that relates to the particle-matrix interaction. Equation (22) could be rewritten as follows:

$$\sigma_{Tred} = \frac{\sigma_T(1 + 2.5\phi)}{\lambda^n(1 - \phi)} = \sigma_{Tm}\,exp(B\phi) \tag{28}$$

where σ_{Tred} is termed reduced tensile strength. From a plot of a natural log-log plot of the reduced tensile strength versus solid volume fraction, the parameter B can be determined.

Rhee [13], investigated mechanical strength of feedstocks (60 vol % iron powder mixed with EVA, polybond and polystyrene) using three point bending technique. It was reported that the strength of a feedstock does not increase by merely increasing the strength of the pure binder. In addition, despite the binder containing EVA has lower strength, the feedstock containing EVA possess the highest yield strength. He attributed it to the higher matrix-powder adhesion for the EVA system due to the polar EVA molecules.

Figure 27 shows the strength of two binders, polyethylene and paraffin based, and those binders loaded with the same volume fraction of carbonyl iron powder. It can be observed that despite the strength of the pure polyethylene binder which is higher than that of paraffin binder, the strength of the feedstock of the former is lower than the latter. The feedstock with paraffin wax exhibits better adhesion to the powder and hence, a greater increase in strength compared to the polyethylene binder. Binder strength

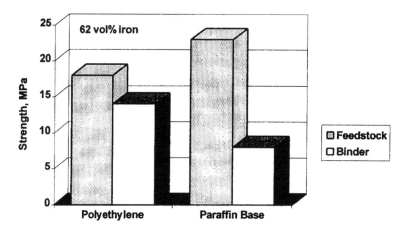

Figure 27.   Srength of two binders and the corresponding feedstocks
(with carbonyl iron powder) [1].

is clearly not a good indicator of feedstock strength.   Instead, strong powder-binder adhesion   contributes significantly to feedstock strength. Findings of Rhee [13] further support the Pukanszky equation (22).

Feedstock strength is also influenced by the powder volume fraction in the feedstock. Rhee [13] investigated using feedstock mixtures of carbonyl iron powder with polyethylene wax based binder. The three point bending test (ASTM 790) was used. Figure 28 shows the result of his investigation. It can be observed that the strength of the compact increases initially with increasing powder concentration until it reaches a maximum at a critical powder volume fraction and then drops afterwards.

This phenomenon   can be explained as a change in fracture mechanism at high powder volume fractions. When the powder volume fraction is below the critical point, after the initial dewetting, the fracture cracks propagate through the binder or the binder-powder interface, depending on whichever is the weaker path. However when powder volume fraction reaches the critical point, the lack of sufficient binder leads to internal voids and agglomeration. The fracture crack is then capable of propagating along the weak points at the particle contacts. Consequently, the feedstock strength decreases dramatically with further increase in powder volume fraction. At this point, it should be emphasized that due to the mechanism of dewetting and reduced load bearing area, powder-binder adhesion is critical. With good powder-binder adhesion, increase in strength is possible at lower powder volume fraction. On the contrary, if powder-binder adhesion is poor, there will usually be a drop in strength with powder volume fraction. Figure 29 shows such an example. As the adhesion between the glass particles and the polypropylene binder is poor, the feedstock strength decreases with increasing powder content.

MORPHOLOGY OF MIM FEEDSTOCK

A morphology study of the feedstock could reveal a wealth of information that could explain phenomenon exhibited during rheological, thermal and mechanical properties study. A SEM (scanning electron microscope) would be more appropriate for this purpose as the small powders normally used in MIM is hardly visible from optical microscopes. Figure 30 shows a SEM micrograph of a tensile test fractured surface of a carbonyl iron powder feedstock. The spherical particles are the carbonyl iron powders. The white grass like structures are the EVA/beeswax binder ripped apart during the tensile failure. To obtain a clearer view of the state of binder-powder interaction, the feedstocks are usually cooled to a temperature below the binder's glass transition temperature so that brittle fracture could take place. This is achieved normally by the use of liquid nitrogen. Figures 31 and 32 show the SEM micrographs of ja cryogenic fractured surface of the carbonyl iron feedstock with EVA/beeswax binder. The powder volume fraction is 0.58. It can be observed that the powder

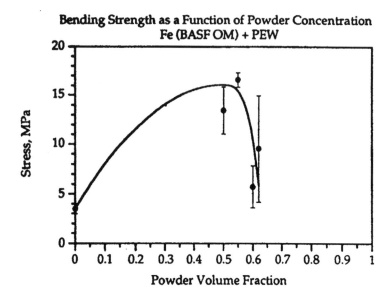

Figure 28.   Effect of powder volume fraction on feedstock
                   strength [13].

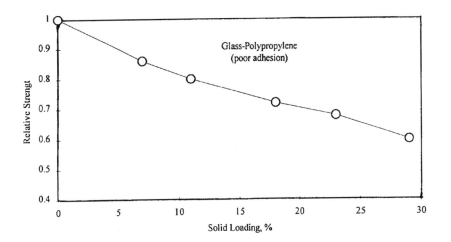

Figure 29.   A plot of relative strength versus solid loading for
a powder-binder mixture exhibiting poor adhesion [1].

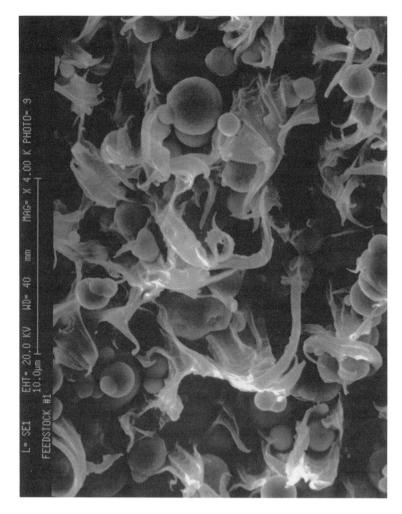

Figure 30.   SEM micrograph of tensile fractured surface of carbonyl iron powder feedstock with EVA/beeswax binder. (X 4,000)

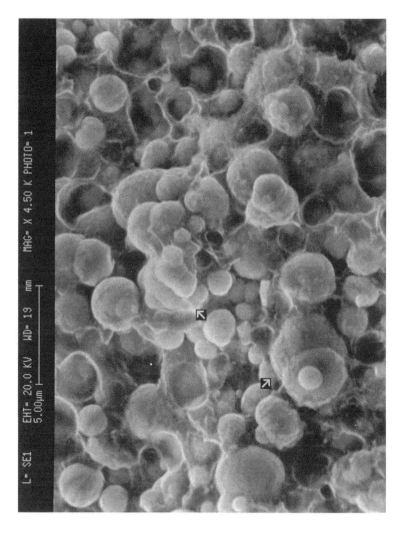

Figure 31. SEM micrograph of cryogenic fractured surface of carbonyl iron powder feedstock with EVA/beeswax binder. (X 4,500).

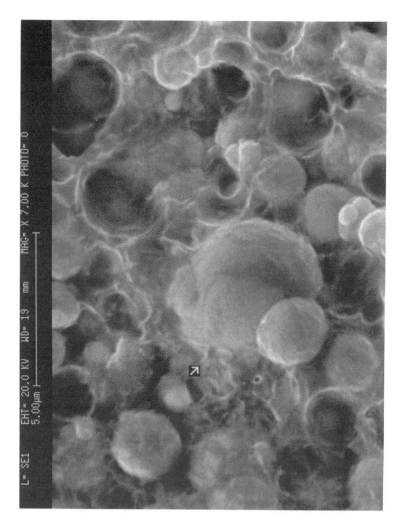

Figure 32. SEM micrograph of cryogenic fractured surface of carbonyl iron powder feedstock with EVA/beeswax binder. (x 7,000)

particles are firmly attached to the binder. No apparent signs of dewetting from the surrounding binder matrix can be found. Agglomeration of the powder particles can be observed very frequently. This is normally due to the difficulty in dispersing the powders especially when the powder volume fraction is very high. The surfaces of the powder particles are found to be covered by a layer of the binder. This could be an indication that the adhesion between the powder and the binder matrix is good.

CONCLUSIONS

The application of polymers to binder formulation for metal injection molding (MIM) has been discussed. The different classifications of binder formulations were listed and described. Rheological, thermal and mechanical properties of the feedstock was described with respect to the effects from powder loading, temperature, binder composition and binder-powder interaction. Morphology inspection by SEM was also discussed. It could reveal the state of adhesion between the powders and binders and is an important technique in the study.

REFERENCES

1. German R.U., Powder Injection Molding, MPIF, Princeton, NJ, (1990), 3-19, 99-120, 134, 147-177, 158, 160, 173, 176,231.
2. Messler, R. W., Jr, Metal Powders Report, May (1990), 363-370.
3. Billiet,R,L, Metal Powders Report, May (1990), 326-332.
4. Messler, R. W., Jr., "Needs and Direction in PIM", in "Proceedings of the PIM International Symposium ", MPIF, RPI, NY, (1990).
5. Chung, C. I, Rhee, B. 0., Cao, M. Y., Liu, C. X., in "Compendium on Metal Injection Molding II", MPIF and APMI, NJ, (1989), 67-78.
6. Strivens, M. A., US Patent, 2,939,199, dated 7 Jun 1960
7. Fanelli, A. J., US Patent, 4,734,237 dated 29 Mar 1988.
8. Sundback, C. A., "Quickset Injection Molding of Ceramic and Metal Materials", in "Proceedings of the PIM International Symposium", MPIF, RPI, NY, (1990).
9. Sundback, C. A., Novich, B. E., Karas, A. E., Adams, R. W., US Patent 5,047,182 dated 10 Sep 1991.
10. Occhionero M. A., Novich, E., Sundback C. A., US Patent 5,047,181 dated 10 Sep 1991.
11. Edirisinghe, M. J. And Evans, J. R. G., J. Mater. Sci, 22, (1987), 269-277.
12. Shah, S. J., "Injection Molding and Rheology of Powder Metal-Filled Compounds for Metal Injection Molding", M.Sc. thesis. University of Lowell, (1988).
13. Rhee, B. 0. , in "Processing Behaviour of Powder/Binder Mixtures in

Powder Injection Molding - Binder Seperation and Quick Freezing",
PhD thesis, Renssalaer Polytechnic Institute, (1992).

14. Rhee, B. 0., and Chung, C. 1., "Injection Molding of Modified Iron
Powder Mixtures with Wax-Based Binders", in "Proceedings of the
Fourth APP Annual Meeting", MPIF, NJ, (1990).

15. Cao, M. Y., Rhee, B. 0., Chung, C. 1., in "Advances in Powder
Metallurgy (1991) Vol. 2, Powder Injection Molding", edited by L.
F. Pease and R. J. Sansoucy, MPIF and APMI, NJ, (1991), 59-73.

16. Stedman, S. J., Evans, J. R. G., Woodthorpe, J., J. Mat. Sci., 25,
(1990), 1833-1841.

17. Zhang, T., and Evans, J. R. G., Journal of the European Ceramic
Society,5, (1989), 165-172.

18. Fergusson, J., and Kemblowski, Z., in "Applied Fluid Rheology".
Elsevier Science Publishers Ltd, England, (1991), 85-136.

19. Cox, W. P., and Merz, E. H., J. Polym. Sci., 28, (1958), 619.

20. Doraiswamy,   D., Tsao, 1. L., Danforth, S. C., Beris, A. N.,
Metzner, A. B., in "Proceedings of Tenth International Congress on
Rheology, Sydney", edited by P. H. T. ULHERR, Society of
Rheology, Sydney,(1988), 300-302.

21. Downey, F., "Rheology and Injection Molding of a Ceramic filled
Polymeric Material", Ph.D. thesis, Comell University, (1994).

22. Isayev, A. 1., Fan, x., J. Mat. Sci., 29, (1994), 2931-2938.

23. Foong, M. L., Tam, K. C., Loh, N. H., J. Mat. Sci., 30, (1995),
3625-3632.

24. Issit, D. A. and James, P. J, Powder Metallurgy, 29, 4, (1986),
259-263.

25. Metzner, A. B., J. Rheo., 29(6), (1985), 739-775.

26. Gupta, R. K., in "Flow and Rheology in Polymer Composites
Manufacturing", edited by S. G. Advani, Elsevier Science Publishers
Ltd., England, (1994), 9-51.

27. Barnes, H. A.,J. Rheo., 33, (1989), 329-366.

28. Hoffman, R. L., in "Theoretical and Applied Rheology", edited by
Moldenaers and R. Keunings, Elsevier Science Publishers Ltd,
England, (1992), 607-609.

29. Nguyen, Q. D. and Boger, D. V., Annual Review of Fluid Mechanics,
24, (1992), 47-88.

30. Lang, E. R., and Rha, C. K., Rha, in "Rheology Volume 2: Fluids",
edited by G. Astarita, G. Marrucci, L. Nicolais, Plenum Press, USA,
(1980), 659-665.

31. Harnett, J. P. and Hu, R. Y. Z., J. Rheol., 33(4), (1989), 671.

32. Dekee, D. and C. F. Chan Man Fong, J. Rheol., 37(4), (1993),
775.

33. Dekee, D. and Durning, C. J., in "Polymer Rheolgy and
Processing", edited by A. A. Collyer   and L. A. Utracki,
Elsevier Science Publishers, England, (1990), 177.

34. Barnes, H. A., in "Theoretical and Applied Rheology", edited by P.
Moldenaers and R. Kuenings,   Elsevier Science Publishers,
NY, (1992), 576-578.

35. Schurz, J., Rheologica Acta, 29, (1990), 170-171.

36. Kalyon, D. M., Yaras, P., Aral, B., Yilmazer, U., J. Rheol., 37(1), (1993), 35.

37. Hansen, P. J. and Williams, M. C., Polym. Eng. Sci., 27(8), (1987), 586.

38. Magnin, A. and Piau, J. M., J. Non-Newtonian Fluid Mech., 23, (1987), 91.

39. Magnin, A. and Piau, J. M., J. Non-Newtonian Fluid Mech., 36, (1990), 85.

40. Windhab, E., in "Proceedings of Tenth International Congress on Rheology, Sydney", edited by P. H. T. Ulherr, Society of Rheology, Sydney, (1988), 372.

41. Yoshimura, A. S., Prud'homme, R. K., Princen, H. M. J. Rheol., 31(8), (1987), 699.

42. Bingham, E. C., in "Fluidity and Plasticity", McGraw-Hill, NY, (1922).

43. Herschel, W. H. and Bulkley, R., Proc. ASTM, 26, (1926), 621-633.

44. Casson, W., in "Rheology of Dispersed Systems", edited by C. C. Mill, London, Pergamon, (1959), 84-104.

45. White, J. L., in "Principles of Polymer Engineering Rheology", John Wiley and Sons Inc., USA, (1990), 209-248.

46. Einstein, A., in "Investigation on the Theory of Brownian Movement", edited by R. Fuerth, Dover Publications, NY, (1956).

47. Fillers, H., Kolloid Z, 97, (1941), 313.

48. Mooney, M., J. Colloid. ScL, 6, (1951), 162.

49. Kupperblatt, G. B., "Wax-Based and Solid Polymer Solution Binders for PIM: Characterisation and Modification", M.Sc. thesis, Renssalaer Polytechnic Institute, (1990).

50. Malkin, A. Y., Adv. Polym. Sci., 96, (1990), 70.

51. Tanaka, H., White, J. L., J. Non-Newtonian Fluid Mech, 7, (1980), 333.

52. Tanaka, H., White, J. L., Polym. Eng. Sci., 20(14), (1980), 949.

53. Theocaris, P. S., in "Metal Filled Polymers: Properties and Applications", edited by S. K. Bhattacharya, Marcel Dekker Inc., NY, (1986), 259-331.

54. Otsubo, Y., Journal of Colloid and Interface Science, II, (1986), 380-386.

55. Song, J. H., and Evans, J. R. G., Processing of Advanced Materials, 3, (1993), 193-198.

56. Rowland, F., Bolas, R., Rothstein, E., and Eirich, F. R., in "Industrial and Engineering Chemistry", 57(9), (1965), 46-52.

57. Evans, J. R. G., and Edirisinghe, M. J., J. Mat. ScL, 26, (1991), 2081-2088.

58. Denizart, O., Vincent, M., Agassant, J. F., J. Mat Sc., 30, (1995), 552-560.

59. Nielsen, L. E., and Landel, R. F., in "Mechanical properties of polymers and composites, 2nd edition, revised and expanded", Marcel

Dekker, NY, (1994), 377-446.

60. Zhuk, A. V., Knunyants, N. N., Oshmyan, V. G., Topolkaraev, V. A., Berlin, A. A., J. Mat. ScL, 28, (1993), 4595-4606.

61. Liang, M. H., Heng, P. C., Chian, K. S., Journal of the Institute of Materials (East Asia), 2, 2, (1993), 53-55.

62. Nielsen, L. E., J. Appl. Polym. ScL, 17, (1973), 3819.

63. Nielsen, L. E., in "Predicting the properties of mixtures", Marcel Dekker, NY, (1978), 49-72.

64. Nishi, T., and Wang, T. T., Macromolecules, 8, (1975), 909.

65. Maiti, P., Chartterjee, J., Rana, D., Nandi, A. K., Polymer, 34, 20, (1993), 4273-4279.

66. Gahleitner, M., Bernreitner, K., Neibl, W., J. Appl. Polym. Sci., 53, (1994), 283-289.

67. Pukanszky, B., and Tudos, F., in "Proceedings of the international symposium on polymers for advanced technologies", edited by Menachem Lewin, Jerusalem, (1987), 793-807.

68. Kerner, E. H., Proc. Phy. Soc. (London), B69, (1956), 732.

69. Halpin, J. C.,J: Compos. Mat. 3, (1969), 732.

70. Chow, T. S., and Penwell, R. C., in "Metal-Filled Polymers", edited by S. K. Bhattacharya, Marcel Dekker, NY, (1986), 227-256.

71. Nielsen, L. E.,J. Appl. Phys., 41, 11, (1970), 4626-4627.

72. Ziegel, K. D., and Romanov, A., J. AppL Polym. Sci., 17, (1973), 1119.

73. Mukhopadhyay, K., Tripathy, D. K., De, S. K, J. Appl. Polym. Sci., 48, (1993), 1089-1103.

74. Lee, B. L., and Nielsen, L. E., J. Polym. Sci.: Polym. Phys. Ed., 15, (1977), 683-692.

75. Pukanszky, B., Belina, K., Rockenbauer, A., Maurer, F. H. J., Composites, 25, 3, (1994), 205-214.

INDEX

X
X-ray diffraction   118

Y
yield stress   234
yield stress   230, 233, 243,
        245, 249
Young's modulus   120

Z
zeolite   187